Technocracy and Development in the Philippines

Technocracy and Development in the Philippines

Roman Dubsky

University of the Philippines Press
Diliman, Quezon City
1993

ISBN 971-542-016-8

Distributed outside the Philippines
by the University of Hawaii Press

Printed in the Philippines by Kaiser Press

Contents

Introduction

This work seeks to explain the rise and influence of technocracy in the developing world. It also examines the causes of this mode of thought, defines its character and its limitations. Technocratic values have not been seriously considered in literature on development, yet the influence of technocratic administrators and planners in developing countries has been extensive. They have had a significant impact on government decision-making and the direction of government policies. This work uses the Philippines as a case study.

The emergence of technocracy is traced to the post-World War II preoccupation with "development". In the Philippines, pressure to "accelerate" the country's development increased greatly in the late 1960s and early 1970s. It is suggested that this emphasis in the new states on "dynamism in development" marks an important stage in the emergence of the authority of technocratic development thought and has led development thought to turn in a technocratic direction. Another theme given prominence is the impact of technocratic values on the recipients of government policies and projects. In spite of commitments made by national leaders to "humanist" ideas of development, there has been an increasing gap between promises of "a better quality of life" and actual performance. This is partly explained by the technocratic thesis. The analysis of technocratic values reveals their counter-humanist nature. This work explores the possibility of overcoming the consequent separation of action and values.

The concept "technocracy" is treated in a broad way, as a system of thought and values, rather than as a description of the actions or influence of particular technocrats or groups of technocrats. The "ideological" dimension of technocracy has been underestimated. Technocracy has been treated as a mere technique used for attaining certain policy objectives which have been defined elsewhere. This view is too simplistic. The collapse of Marxism as a viable state ideology in Soviet Russia and other communist-dominated regimes in the late 1980s and early 1990s draws attention to the fragility of communist social and political models. At the same time capitalist-pluralist economic and political systems continue to display their perennial tendency towards instability and inequality, as illustrated by the deepening recession of the early 1990s with the distressing consequences for wage-earners and other vulnerable groups. These problems in existing models may stimulate a quest for ideas more fitted to the present-day situation, and " technocracy" may offer itself as a viable alternative. This is because its approach is perceived to be based on expert knowledge and science and to allow objective judgement in decision-making; it is also associated with the popular and appealing idea of productive efficiency. When contrasted with existing systems, it suggests the possibility of a rationally ordered social system. Seen from this perspective, technocracy assumes an ideological position, which, like all established ideologies, has a universal message. Indeed, the possibility—some say the danger—of this happening is one of the topics of this work.

I have given prominence to two concepts: technocracy and development. They are interconnected, for technocratic thought is defined as an aspect of development thought. " Technocracy" is the central concept. This work seeks to establish the validity of the technocratic thesis by surveying government policies and Marcos's writings for the presence of technocratic elements. It also contrasts the early " liberal phase" of development of the 1950s and 1960s (Chapter 2) with the " dynamism" of the 1970s with its preference for technocratic approaches in policy-making (Chapter 3). In addition it shows that technocracy also continued to play a central role in Philippine policy-making after the demise of the Marcos regime.

"Development" was the central ideological concept during the term of President Ferdinand Marcos for describing the idea of national progress, and its ideological utility was appreciated also in the subsequent era. Unlike the notion of the New Society, which, after, Marcos's demise, was excluded from the vocabulary of Philippine politics, this concept retained its vitality, unlocking sympathetic feelings among both national planners and leaders and the purses of foreign-aid donors.

Although this work covers a variety of topics, its main objective is limited. It does not seek to describe or evaluate the state of social and economic progress in the Philippines between 1950 and 1980 or the contribution of the New Society to Philippine development under Marcos. Rather, it seeks to highlight the rise and influence of technocratic thought in government policies and planning, and to identify adverse effects of the technocratic position and consider the possibility of reducing them.

A few observations can be added to explain this work. First, it follows the format of the dissertation upon which it is based. However some material has been added. This includes a section on " Technocracy and Freedom", which is intended to highlight the danger to freedom from technocracy-dominated systems, and a chapter that briefly covers the regime of President Corazon Aquino. The original introductory chapter, which deals with methodological aspects and includes a review of the literature, has been retained under the title " Technocracy and Development", for it provides useful material to students in the field. It may, however, be omitted by those readers who have no patience with such " background" material without impairing understanding of the overall argument of the work.

Some changes have been introduced in the text which reorient aspects of the original thesis. The reference to the " new directions" in the development of the Marcos regime (in contrast to the " old directions" of the preceding liberal era) has been dropped in favor of " dynamism" of development. This is because the former concept appears excessively value-loaded and because the latter suggests the prevailing thrust towards " accelerated growth" and has more direct technocratic implications. In retrospect, many reforms presented as radical at the time of their appearance did not lead to a major break with the past.

The "New Society" did not take firm root. Marcos's economic reforms, based on a centralization of decision-making and comprehensive planning, utilized new models and strategies for dealing with the country's economy, but their success was ephemeral. In the latter part of the Marcos regime the economy experienced stagnation and crisis. Other "changes" were frequently no more than political rhetoric—a familiar feature of the Philippine political tradition. Often the style underwent change, but the basic structures continued to exist, assuming new forms or names. The most lasting contribution of the Marcos years appears to have been the process of "technocratization" of the government system, although some observers of Philippine politics and development would view this as a mixed blessing at best.

Help was provided in the writing of the original dissertation by a variety of people. These included my adviser Dr. Ledivina Cariño from the College of Public Administration, University of the Philippines, who made helpful suggestions about the focus of the study and a number of important points of definition. Useful assistance was offered by Dr. Estrella D. Solidum from the Department of Political Science, the University of the Philippines, who also kept my at times flagging interest alive through discussions on suitable definitions, focus and other aspects. I am most grateful for their help. I also wish to express my gratitude to Mr. Richard Dorrell, who made several documents available to me during the writing of the study. In addition, I wish to thank my friend Jun Garces, who looked after other aspects of the work, particularly in its last stage, arranging its printing in my absence from the country—not a small task if one knows the complexity that often arises in these matters in a developing country.

Outside the Philippines, I am grateful to Dr. David Routledge from the University of the South Pacific for valuable suggestions with regard to the "humanist" theme of the study. Mr. Alan Robson also helped by raising a number of provocative and critical questions about the nature of the present argument. I also wish to express my appreciation to the University of the Philippines Press for making this publication possible.

ROMAN DUBSKY

Technocracy and Development in the Philippines

CHAPTER I

Technocracy and Development

The theme of this work is the rise and growing influence of technocracy in developing countries. It encompasses the original contribution of technocracy to social and organizational thought. It highlights the fact that technocracy has generated several fresh and influential ideas and approaches that have turned human thought and action in a new direction. In addition the work highlights the impact of technocracy on existing ideas of development as well as on people affected by development policies and projects, and identifies the limitations of this position. The work concentrates on the values or beliefs, the "ideology" that characterizes the technocratic position rather than on actions of particular technocrats or groups of technocrats. It draws on the experience of government systems rather than the private sector, and takes one developing country, the Philippines, to illustrate this experience. The opening chapter concentrates on definitional and methodological aspects and includes a historical background.

The origin of technocracy can be traced to the post-World War II years. Technocratic influence at first spread in advanced industrial countries, but later developing countries followed suit. It did not take a long time for ideas propagated by technocrats to become more widely recognized; they proved to be very useful because of their relevance to their times. They could easily be attached to a number of concepts that were becoming dominant in the context of change, such as "development," "modernization," and "productive efficiency." Technocratic ideas also gained attention due to their association with technology and science.

Paradoxically, despite the growing influence of technocracy, there is a paucity of works written on it both in developed and developing countries.[1] Although references to it appear in many works on modern organizations, administration, policy-making and development, the level of analysis is frequently low, and the composition of technocratic values is not adequately explained. Moveover, technocracy tends to be treated as a phenomenon characteristic of advanced industrial systems, and its growing importance in the context of development is neglected. The objective of the present work is to partly remedy this situation.

This work has two objectives. The principal aim is to describe and explain the emergence of the authority—the diffusion and penetration—of technocratic thought in the context of development as well as to draw attention to the increasing orientation of development ideas in a technocratic direction. This process is identified as it manifested itself in the behavior of government systems and policies in the Philippines over the period from 1950 to the 1980s. Emphasis is given to changes that occurred during Marcos's martial law regime and they are traced to attempts to achieve rapid growth in the economy. It will be suggested that the emphasis on "dynamism" in development marks an important stage in the emergence of the authority of technocratic thought. The term "dynamism" will be taken as a characteristic of actions undertaken by policy-makers to accelerate the process of development by enhancing its comprehensiveness, intensity and efficiency; "the authority of technocratic thought" refers to the role of ideas in impelling action in a technocratic direction by effecting a corresponding change in the perspective of development thought and by establishing its extensive presence in development activities. This discussion also aims to define the extent of technocratic influence and to identify the factors that have given rise to it.

The second aim is critical: to draw attention to the weakness of technocracy in its impact on people affected by development policies and projects. In technocratic planning, concerns with individual and social development—sometimes referred here as the "humanist" dimension in development—tend to be neglected or are expressed in terms that rob them of their meaning. In the Philippines, despite the government's perennial commitment to a "better quality of life" for everyone, "humanist" objectives have not been achieved. This seems to have been to a large extent due to the technocratic component that went into government policy planning and projects. Planners looked for technical solutions to people's problems which were popularly perceived in terms of particular needs. Thus their respective perceptions often differed, reflecting the inability of technocratic officials to accommodate people's immediate development needs into their policy design. This conflict will be illustrated by several case studies. The possibility of overcoming or reducing the dualism of values implied in this conflict will be considered in the concluding chapter.

In a more general way, this work seeks to draw attention to the emergence of technocracy as a form of ideology and its growing influence in crucial areas of decision-making. The importance of this influence has not been adequately appreciated, yet the emergence of technocracy charts a major shift, if not revolution, in social thought and values. A partial reason for this may be the relatively low profile of technocrats in public life, shrouded as they are in bureaucratic anonymity. Another reason seems to be that traditional ideologues do not feel threatened by technocratic ideas as much as by competing traditional beliefs. At least they seem to view technocratic ideas as a lesser evil. This reaction may, however, be based on a simplistic understanding of the powerful attraction that technocratic approaches may have for industrial and post-industrial systems. It seems to overestimate the influence on them of traditional social and political philosophies and to underestimate their interest in social change towards a technology and science-based outlook.

There are a number of reasons why the influence of technocracy has not been given adequate cognizance. First, the existence of technocracy as a *universal phenomenon* has not been sufficiently recognized. Technocracy tends to be identified with capitalist countries.

Yet the same ideas and practices are followed in non-capitalist systems. The management of the centralized *Gosplan* state-planning system in former Soviet Russia is an excellent example of the role of technocrats in economic policy-making in a communist country. They exercised their influence through administrative means and got support from technocracy-minded politicians; most of them were members of the Communist Party. Indeed, many people hailed Mikhail Gorbachev's ascent to power as the rise of a new Technocratic Age moving the country towards a more rational system of social and economic development. Highlighting a similar point, Professor John Kenneth Galbraith says that " convergence between the two ostensibly different industrial systems [the Soviet and the American system] occurs at all fundamental points." In his view this is " an exceedingly fortunate thing. In time . . . it will dispose of the notion of inevitable conflict based on irreconcilable difference."[2] Such a convergence may be accelerated with the dismantling of the Soviet Union into its constituent republics.

Second, the extent of technocratic influence in policy-making has not been adequately appreciated. This influence has been largely carried over from science and organization theory; their criteria tend to be used in evaluating social issues and policies. This point has been made by many social theorists (e.g. Mannheim), that there is a tendency in modern systems to evaluate social or policy issues by criteria of technology and science and to translate such issues into the language of technical " problem solving" rather than in more orthodox terms. According to them, this applies both to " capitalist" and " communist" systems, blurring the distinction between them, and they attribute it to the " imperatives of technology and organization, [which, rather than] the images of ideology . . . determine the shape of economic society."[3]

Third, the attraction of technocracy as a model for the construction of social systems has not been seriously considered. Yet the worldwide collapse in the early 1990s of so many communist regimes shows the basic fragility of Marxist-based social systems, although the unfettered capitalist alternative is regarded by many as equally undesirable. This may lead to a search for a new social model, which combines scientific approaches with the idea of the welfare state as it has evolved in the West. In this context, technocracy with its reliance on advanced technology and science may be seen as a technically efficient and "ideologically neutral" ("objective," "impersonal," and "impartial" like science) tool for achieving an alternative social and political development.

Fourth, the influence of technocracy is viewed too narrowly, as confined essentially to planning. But, as this work demonstrates, its influence is widespread and permeates the whole government administrative and decision-making system. Arguably, when the decision-making process becomes excessively technocratic, even strongmen like Marcos may find themselves in agreement with their technocratic planners, for the success of their political programs and, ultimately, their capacity to stay in power will depend, to a large extent, on the planners' performance. Another example is the increased investment in most developing countries in technical education as against traditional liberal-arts education. This suggests their recognition of the importance of technology and science for accelerating development and is likely to strengthen the influence of technocratic approaches and values.

Fifth, the importance of technocracy tends to be viewed only as a means, while traditional ideologies represent social and political ends, suggesting the direction society ought to take. Arguably, technocracy can attach itself to any ideological system. However, it can also be viewed as representing its own particular vision of ideal reality. This suggests a rather technical ideal on which to construct a new social order, but, as will be seen, the present technocratic goals and approaches can be modified to incorporate ideas of human development which conform more closely to the classical notions of the Good Life. Because of its solid roots in technology and science and their almost universal appeal, the potential of technocracy as a formula for well-ordered social systems should not be underrated.

A note delimiting the range of this work is in order at this point. First, it concentrates on the emergence, the nature, the influence and the limitation of technocratic ideas and the impact of technocratic policies and projects. It does not consider the behavior of particular technocrats or groups of technocrats. Also it does not deal with "the politics of technocracy," although it briefly describes policy debates and power conflicts in which Philippine technocrats played a prominent role. The appreciation of power-struggles involving government technocrats is obviously very important for an understanding of the social and political forces that shaped the policy-making process in the Philippines, but this work has a different orientation. Second, in surveying Philippine development thought, this work confines itself only to the study of government systems. It is not intended to be a general review of postwar development thought. The historical treatment of the argument has been undertaken to provide a contrast between the "liberal framework" of the Philippine polity and the later emphasis on accelerated development. Third, this work draws mainly on government documents and Marcos's writings. This makes it vulnerable to criticism stemming from the use of unreliable sources and government propaganda. In this connection there is some difficulty in interpreting Marcos's writings. The gap between what is found in his "ideological" works and what was happening in the country is often so wide that it is not easy to say whether Marcos's ideology was a mere exercise in hypocrisy or a meaningful statement of a particular brand of nationalist idealism. Despite these difficulties, however, there are consistencies both in government activities and Marcos's actions and writings that should allow us to generalize about the government's philosophy of development.

The Historical Background

This work draws on history and the literature in the field of technocracy. These have a variety of definitional aspects. A " technocratic orientation " in social policy can be traced to 19th-century thought and practice and was essentially a response to the issues raised by the Industrial Revolution. The founding father of this orientation was the French aristocrat Saint Simon, a prophet of the new Industrial Age, who, like Marx, saw its enormous potential for facilitating progress. Science, applied to industry and properly managed, was expected to lead to the efficient use of physical and human resources without waste. Thus, by rational management and planning, social evils of the past—the political rivalries that breed wars and the economic inequalities at the root of human poverty—would be eradicated. Saint Simon's vision of an ideal social order was a system of organizations guided by industrial experts; on the spiritual side, guidance was to pass from the church to the men of science. The impulse underlying this early manifestation of the technocratic spirit was humanitarian and the various schemes for social amelioration embodied a dose of authoritarian planning, anticipating the managed societies of a later era. From its beginning the technocratic idea located itself in industrial thought and practice.

Technocratic elements may be found in many 19th-century European socialist schemes. These assailed poverty and wastage of resources and promoted social engineering. The scientific-technocratic idea also gained support from early modern sociology. Auguste Comte came to view the "positivist" (scientific) stage of evolution as the acme of human progress. In his later thought Comte sought to establish a new "Religion of Humanity" based on his positivist ideas of science.

The technocratic orientation is frequently associated with social engineering, the idea of "skillful management or contrivance" (Webster) by means of some rational or scientific scheme in order to restructure a social system. An early example of this is the utopian scheme of Charles Fourier's *phalansteres*, or producers' settlements, which are both working units and a community of persons living together in a single great building. Fourier implied that persons who could be made to fit into such a socio-economic arrangement would lead the best and most productive existence. Almost all utopian social schemes have such social engineering components.

Technocratic and social engineering outlooks have been closely linked in the various schemes of social reform since the 19th century, as social engineering has been increasingly viewed as a useful method for dealing with the consequences of the Industrial Revolution and the Demographic Revolution. Capitalist expansion was responsible for growing poverty, increased ill health in industrial areas, rapid growth in population and a shift to the cities. The old system of social security, based in England on rural parishes, collapsed, drawing attention to the need for some measure of innovative social planning. Problems arising in the big cities were becoming similarly unmanageable. As a practical formula for public policy, the social engineering approach is reflected, for instance, in the economic theories of the English economist Keynes and his followers; it is identified with state economic interventionist and regulatory strategies in the context of the market. It is also reflected in the concept of the Welfare State to the extent that social welfare is introduced by means of planning "from above."

The strongest influence of the technocratic orientation was in America. This was due to a number of factors. One of these was the emergence of scientific management associated with Frederick Taylor in the early decades of the century. Taylor, an engineer by profession, developed a number of theories to show that there is "one best way" that can be objectively established for the effective operation of industrial systems. Even more influential were the sociological theories of Thorstein Veblen, an academic of unorthodox views, who, at the height of *laissez faire* in America, was predicting the impending doom of this system. As he saw it, the interest of business and finance capitalists was increasingly coming in conflict with the system of mass production, for businessmen were after profit and did not really care about production as such. Veblen portrayed the successful business class in Darwinian terms as the last historical manifestation of man's primitive "predatory instinct," which, in his view, made a rationally ordered society impossible. Underlying his revolutionary challenge to capitalism was his worship of industrial progress and industrial rationality. Indeed, today's technocracy may be said to have been born in a spirit of intellectual revolution, which, like Veblen, was inspired by a vision of a rationally ordered society, based on industry and full use of productive resources.

An American work with a more explicit technocratic orientation appeared in 1934 under the title *Technocracy Study Course.* This was composed by a group calling itself Technocracy Movement, later Technocracy, Inc. Headed by a charismatic engineer named Howard Scott, these friends of technology evangelized about the benefits of scientific rationality, which, during the depression, seemed to offer a non-revolutionary way to salvation. The work was not only a practical primer but also preached the beauty of rational living based on science. The principles of scientific control were to be applied to all spheres of existence, including the management of society and economic systems. In the words of its contributors, the work was to remind the American public of their "rendezvous with Destiny" and offer them "the blueprints of Continental [i.e. all-American] physical operations for the production of abundance for every citizen." They insisted that this "can only be accomplished by a Continental technological control—a governance of function—a Technate."[4] This group became increasingly inactive with the lifting of the depression in America and the launching of a number of progressive development-generating programs under President Roosevelt's New Deal policies (themselves perhaps the earliest example of state-initiated social engineering in America). Its *Course* survives as a literary curiosity.

After World War II, American interest in technocracy was associated with the problem of how to manage vast industrial and post-industrial social systems. A number of well-known works were published. Writing from the perspective of a post-industrial age, some, like Roszak, treated technocracy critically, while others, like Daniel Bell, sought to explain and justify its existence. For the critical writers, technocracy, because of its overemphasis on science and techniques, is a social evil and they condemned it harshly as dehumanizing; in the phrase of one such critic, "The invasion of technique de-sacralizes the world in which man is called upon to live."[5] On the other hand, Daniel Bell viewed the technocratic mode of existence as an inevitable consequence of changes in social structures that are occurring in advanced societies. According to him, a new type of society is coming into existence which is innovative and is characterized by a change from empirical and trial-and-error modes of thinking to "theory and the codification of theoretical knowledge for directing innovation and the formulation of policy."[6] He described the "basic" existential problem of such post-industrial or technology-based society as follows:

> Every modern society now lives by innovation and the social control of change, and tries to anticipate the future in order to plan ahead. This commitment to social control introduces the need for planning and forecasting into society. It is the altered awareness of the nature of innovation that makes theoretical knowledge so crucial.[7]

A more sober argument for today's acceptance of technocracy was developed by Galbraith in his work *The New Industrial State.* Galbraith uses the term "technostructure" to describe the new industrial system which, he believes, is the dominant force in the contemporary world. This is based on the belief that large polyglot corporations using advanced technology and science, are closely "fused" with the state and dominate economic systems "by extensively replacing market with planning."[8] This system is justified by the argument that since in modern productive systems a lot of time, capital and other resources have to be committed to production, there is a need to carefully control and manipulate supply and demand to minimize risk, and this is done by comprehensive planning. The goals of this system are economic: "the expansion of output, the companion increase in consumption, technological advance, the public images that sustain it . . . [and] all of our lives will be in the service of these goals."[9] The state increasingly identifies itself with these goals and systems, often treating such organizations as an extension of its own bureaucratic system. In Galbraith's words, " the line between the industrial system and the state becomes increasingly artificial and indistinct."[10] Hence, " the imperatives of technology and organization . . . are what determine the shape of economic society," and Galbraith warns that this trend should be resisted if we are not to become " the servants in thought, as in action, of the machine we have created to serve us... [because] what counts is the quality of life."[11] For him " the educational and scientific estate" and the intellectuals should provide the necessary political and moral inspiration and " the right social purpose."[12]

The tradition of technocracy is also strong in France, where technocratic thought has its longest history, stretching back to the early years of the 19th century (Saint Simon). In French organization thought the formal and disciplined system of the French civil service, which extends back to Emperor Bonaparte's administrative reforms, has become a carefully cultivated legend through education and recruitment requirements. In education, it has been associated particularly with two elite institutions, the *Ecole Polytechnique* (for engineers) and the *Ecole Nationale d'Administration* (for administrators), which provide professional and technical training for top recruits. The French literature contains some of the best works written on technocracy. The most prominent of these is Jean Meynaud's classic study *Technocracy*, which describes the administrative elite of the French state bureaucracy. The work offers excellent insights into the dynamics of technocratic activity, but its principal emphasis is on technocratic domination and power. The rise of technocratic power is perceived to be at the expense of political power, as if the misfortune of the latter were the fortune of the former. Technocracy tends to be equated with the power exercised only by senior officials, which makes the concept somewhat narrow; nowadays the technocratic influence appears to be diffused more widely. Another French writer, Alain Touraine, addresses himself, like Daniel Bell, to the " post-industrial" social order.[13] He describes this in terms of a "programmed society" and explicitly recognizes the technical expert as the principal moving force behind it. Technocrats are the big managers and manipulators who keep the system going, well integrated and well planned. However, they have no political ambitions: "Their ideology is service to the State, to the party in power, to the economy; their ethics is impersonal; their style is manipulation rather than imperiousness."[14] More specifically, their special task is to look after "growth and development" and "manage" society in that direction; they consider "society to be only the collection of the social means to be used to achieve growth and to reinforce the ruling structures that control it."[15]

In England the technocratic orientation is associated with the tradition of social reforms, and explicit prophets of technocracy of the French kind are lacking. It was implicit in the influential theories of the early 19th century utilitarians, particularly Jeremy Bentham and his successors, who advocated radical reforms in politics, administration and ethics. Inspired by Newtonian determinism, Bentham sought to discover scientific laws of behavior which could be applied to social and legal reforms. The core of his philosophy was the principle of utility and self-interest. This was based on a mechanically conceived psychology, in which human behavior is interpreted in terms of the inclination to seek pleasure and to avoid pain. Utility was to be measured objectively as amounts of these sensations. Bentham's rational outlook and his practical, non-sentimental, materialistic system has an affinity to many ideas characterizing the technocratic outlook. Through the utilitarians these ideas were adopted in England, becoming part of an established tradition of social thought which encouraged social and political reforms in accordance with the principle of rational utility. A similar spirit dominated a great deal of English Fabianism with its zeal for utopian social reforms backed up by an inclination for rational, vaguely authoritarian means to bring them about. Focused explicitly on technocracy is a work by Peter Self entitled *Econocrats and the Policy Process.* It seeks to identify the social and political assumptions that underlie models used by economic planners and subjects to criticism welfare-economics theories and techniques used in policy-making, particularly that of cost-benefit analysis.

The argument of James Burnham, an American social scientist, belongs to a different category. It concentrates on the political power of technocratic managers in modern political systems. In his *Managerial Revolution*, published in 1941, Burnham predicts that a managerial type of political organization is becoming universal and inevitable because of the obsession of modern organizational systems with productivity and efficiency. He equates this form of organization with comprehensively planned and strongly centralized social and economic systems where the means of production are controlled by "managers" (the original capitalist owners become increasingly irrelevant); these are technicians, bureaucrats and business executives. Executive power is concentrated in their hands, and in politics promotes oligarchic regimes through which they dominate the population at large, whether by fraud (getting people's support by utopian promises of a New Society to come), or by force. This system is neither capitalist nor socialist in the traditional sense; the old capitalist class is eliminated and the working class is reduced to an essentially servile role. Burnham contends that such systems can retain their dynamism under certain conditions. In his later book, *The Machiavellians*, published in 1943, Burnham refines his argument by suggesting that in order to retain power, which is its real aim, the ruling technocratic class should not be narrow and selfish, but rather should serve the common good and make its membership accessible to other people. He emphasizes the usefulness of Pareto's theory of the "circulation of the elites" for retaining dynamism as well as power in technocratic systems.

A number of the well-known fictional works from this period dwell on the theme of social manipulation. In Aldous Huxley's *Brave New World* (1932), the purity of the system is ensured by the mass production of babies in test tubes and the use of conditioning techniques and drugs to shape the citizens' behavior in order to achieve a painless and happy existence for everyone. George

Orwell's *Nineteen-Eighty-Four* (1949) is a vision of the Administrative State in which all aspects of life are controlled by a totalitarian State, the omnipotent "Big Brother," to ensure both physical and emotional conformity. Both works draw attention to the impact of excessive social manipulation on human freedom. Manipulation is also the theme of two novels by B.F. Skinner, *Walden Two* (1948) and *Beyond Freedom and Dignity* (1971). Unlike Huxley and Orwell, Skinner welcomes social manipulation through technocratic means as a scientific method for attaining human happiness. Both works are based on Skinner's psychological principles according to which persons are a product of the environment and reinforced stimuli. Positive reinforcement by rewards leads to repetitive behavior; correct reinforcements can be used to improve society. Such a science-based society is illustrated in *Walden Two*: here government is operated by experts; all citizens live in one large building, and behaviorist conditioning principles are practised. All citizens have to work but "dirty work" is made more attractive by earning the citizen more credit. This is to ensure, by using such positive reinforcements, that dirty work gets done, while at the same time the citizens are free to choose their work and leisure activities. They obey and participate in community objectives because they are positively reinforced for doing so.

Social engineering relating specifically to the Philippines was the subject of a book by G.L. May entitled *Social Engineering in the Philippines*, published in 1980. According to May the phrase "social engineering" is used "to describe an effort to mold, and often to restructure, a society."[16] The work describes the main policy strategies—education, economic reforms and preparation of Filipinos for self-government—used by the American colonial administration in the Philippines during the period 1900–1913 to create a more socially and politically advanced system. However, the definition of social engineering is not explained further. Indeed, it appears to fall outside the usual meaning of the concept. It misses the engineering-methods aspect with its linkage to the notion of "scientific control" as well as the emphasis on an ideal social order.

In fact, in developing countries the contribution to the literature of technocracy has been very limited. It is confined almost exclusively to learned articles or journalistic writings. In the Philippines, for example, the emergent technocracy of the 1960s and early 1970s is identified by Professor Romeo Ocampo in a study published in 1971. For Ocampo the technocratic trend is represented by a group of American-trained economists who assume the role of idealist modernizers eager to apply their recently acquired skills to turn the national economy in a more dynamic direction. An empirically based systematic study entitled *Technocrats as Middlemen and their Networks in the Philippines* has been written by Victoria M. Arcega. The author correctly identifies technocratic influence at different levels of administrative systems. Her particular interest, however, is technocratic culture. In her concept of "personal networks" she highlights certain affinities in behavior and feelings among technically oriented administrators, which for her suggests a fairly identifiable form of "group status" and indicates "the possible emergence of a new type of cultural heritage and new social patterns among this group of people."[17] She refers to this group as "the third culture" and defines its task as one of mediating new ideas, whether between science and bureaucracy or bureaucracy and the public. They serve as brokers or "men-in-the-middle" to interconnect two or more segments of society or, to put it differently, to "synthesize innovations of a technical type, mediate the research-scientific circle (first culture) and the policy-makers (second culture)."[18] Arcega also draws attention to the adaptability of this group. In their tasks of brokerage, they tend to accommodate themselves to prevailing cultural values while maintaining their particular orientation. "The technocratic values are not absent in these transactions," she says, "but they are normally welded with preferred norms of Philippine conduct."[19]

Defining Technocracy and Development

The main concepts used in this work have to be defined; these include " technocracy" and " development" and " humanist" ideas of development. There is no standard definition of technocracy in the rather thin literature on this subject. However, the term consists of two roots, " techno" and " cracy," which suggest its orientation. The former is related to the Platonic concept of " the possession of a craft or skill to perform to perfection"; the latter is based on the Greek work for " government." Thus a literal translation is: " government by experts," which suggests the origin of technocracy in an industrial age. A more specific definition is given in the Oxford English Dictionary where technocracy is treated as a social system and is defined as " (the state where there is) organization and management of a country's industrial resources by technical experts." The present work seeks to define the philosophy or " ideology" underlying technocratic behavior and systems. " Technocratic thought" may be defined as a set of ideas which aim to enhance the rationality and effectiveness of systems in which the appropriateness of actions is judged essentially by technical or professional criteria and where systematic use is made of advanced methodology of science.

Viewed more broadly, the technocratic orientation can be defined in terms of a cluster of attributes. A survey of literature in this field suggests five of these:

1. Technical orientation, expertise. Technocracy has been called " government by technique" (Gournay) or, in Roszak's phrase, a " regime of experts . . . who justify themselves by appeal to scientific forms of knowledge... [for whom] beyond the authority of science there is no appeal."[20] Technocrats are said to have " impenetrable skills" (Roszak). The technical thesis must be, however, qualified. Expertise or technical knowledge by itself does not turn people into technocrats. Furthermore, technically oriented organizations or systems may or may not be

technocratic; such organizations and technical experts existed even before the advent of "technocracy." It is not, then, mere presence of technical factors that makes organizations or people technocratic; rather, it is an *excessive* emphasis on such factors. This point is made by Peter Self: "Technocracy should refer to the excessive power or influence of technical experts resulting either from an exaggerated weight being put upon some technical factor in a policy decision, or from the aggrandizement of the function with which the expert is particularly concerned."[21]

2. An emphasis on planning, control, predictability, bigness. Technocracy can be viewed as "an organization for the control of industrial resources by technologists" (Oxford English Dictionary). Planning, forecasting and methods of control are said to be necessary for the successful management of systems to maximize their operational certainty, and hence their predictability and efficiency. Methods used for control must, however, be big in scope to be effective. Hence there is an emphasis on the integration of systems. Advanced techniques of social science such as systems analysis, linear programming and program budgeting, and rational macro-models are used in decision-making.

3. A managerial outlook. Technocrats can be said to assume the qualities of "managers," taking, in Meynaud's words, "an over-all view, much broader than that taken by a mere technician or expert."[22] Similarly, according to Touraine, they "are not technicians but managers, whether they belong to the administration of the State or to big businesses."[23] They represent particular professions where management skills are appreciated, working, in Dreitzel's words, "as science administrators, planners, coordinators and advisers."[24] Although they are sometimes defined as "organized technologists" (Oxford English Dictionary), they should be perhaps differentiated from them because the technocratic outlook shows preference for macroscopic approaches, while technologists tend to be associated with methods of engineer-

ing, characterized by the use of microscopic approaches and problem-solving activity.[25]

4. An ostensibly "apolitical" outlook. Technocracy is held to be "ideologically invisible," constituting, according to Roszak, "a trans-political phenomenon following the dictates of industrial efficiency, rationality and necessity."[26] Not unlike Weberian bureaucrats, technocrats perceive development in terms of production, not political forces. Their preference for political neutrality must be, however, qualified. It means that they employ technical rather than political criteria when they exercise their judgment, but this does not exclude them from policy-making. Indeed, they tend to be actively involved in it and "articulate the scientific-technical culture with policy-makers."[27]

5. A world-view centered on physical and material rather than spiritual, aesthetic or religious values. Bell describes it as a "logical, practical, problem-solving, instrumental, orderly and disciplined approach to objectives, in its reliance on calculus, on precision and measurement and a concept of a system."[28]

Technocrats have the following characteristics: (1) They are persons equipped with some specialized competence or professional skills which are the principal reason for their employment. (2) Their skills ("technocratic skills") are of a particular kind, applied mainly to problems of organizational management and planning and include the rationalization of systems. They are confined mainly to the area of economics, business and public administration, social engineering and planning (e.g. national, environmental, local, in production and marketing) and are identified with the tools used in these areas. It may be noted that not all professionals, economists or engineers would qualify as technocrats under this definition. For instance, engineers may be employed in essentially micro and action-oriented activities rather than in positions requiring managerial skills and planning. Furthermore, "pure scientists" are not usually involved in "technocratic activity." Such professionals as physicians or lawyers *qua* practitioners of their profession would also be excluded. (3) Their position involves a degree of access to decision-making and of discretion; this may be exercised at different levels, including local decisions and policy implementation. (4) They are usually college-educated persons or possess comparable qualifications. (5) They are found both in public and private employment. (6) Because of their peculiar skills, training and outlook, they should be differentiated from conventional officials, although this distinction may be blurred. This is because the bureaucrats who have been exposed to technocratic values seem to have acquired something of a "technocratic outlook," and with increased technocratization of administrative systems such an outlook may become more extensive and (to use Max Weber's description of bureaucratic conduct) second nature to them. (7) It may be anticipated that with the progressive technocratization of organizational systems, the notion of a technocrat will shed its exclusiveness and will be more commonly applied to decision-making activities at all levels of these systems. However, the basic technocratic attributes defined earlier are likely to persist as distinguishing characteristics of technocracy-oriented systems.

The concept of development in this context will be treated as an essentially ideological one and will be identified with "modernization." In the Philippines the notions of development were tied to modernization theories of social and political progress. They were imported into the country at an early stage, in the wake of the establishment of the Philippine Republic in 1946, and followed American development models especially strongly because of former colonial connections. Modernization theories lay emphasis on technology, industrialization and economic factors and dealt with such trends as urbanization, agrarian problems, population growth, increased literacy, political development and other attributes of "modernity." In these theories "development" is defined as a process of change—in particular, the transformation of social and political systems in accordance with essentially Western ideas of human progress. It is conceived as a process that is directed toward some end, typically the goals stated in national development plans. Used as a formula for progress, the notion of modernization has a strong ideological dimension. The past successes of industrial capitalism are universalized and their type of development becomes the ultimate goal for all development. Thus the concept is used not only descriptively, to identify behavioral tendencies in this direction, but also prescriptively, to serve as a model or formula for action. Some writers, like W.W. Rostow, point to the historical inevitability of this process. Rostow contends that studies of history reveal several stages of development and that these stages have "an inner logic and continuity" of their own and suggest the possibility of a universal law that constitutes "both a theory about economic growth and a more general, if still highly partial, theory about modern history as a whole."[29]

"Humanist" ideas of development refer, in the context of this work, to notions that highlight the commitments of policy-makers to "humanist" values. Like technocratic values, humanist values are one aspect of development thought, although they predate the latter. Humanist notions of development emphasize well-being in terms of *personal* improvement in the "quality of life" and show particular concern for social welfare, basic human needs and similar objectives aimed at encouraging people's physical and intellectual self-development. Humanism implies a "holistic" approach to development in which there are organic and functional relations between parts and wholes. Contemporary references to this outlook are found in the rhetoric of political leaders in developing states, the U.N. commitment to "development decades," which include "social development" and "social justice" among their principal goals, and the growing literature criticizing the inadequacy of prevailing notions of development. Such pronouncements highlight the idea that social or humanist concerns and economic concerns should be viewed as interrelated. This notion is, however, problematic. Although references are made to it in Philippine writing on development, indeed by President Marcos himself, it belongs to the realm of political rhetoric rather than action. Moreover, the experience based on the conduct of Philippine planners dealing with issues of social development suggests a return to a more "paternalistic" impulse as the "social sectors" become subject to planning and control from above by a "benevolent" state.

The Framework and Methods

This work explains the emergence of technocracy and its influence as a consequence of the commitment of Philippine leaders to national progress through modernization. As early as the mid-1940s they decided to adopt the modernization model as the guiding orientation for the country's social and economic policy. This model, brought into the Philippines, as elsewhere, mainly through American influence, laid emphasis on technology and industrialization. It encouraged the adoption and use of technological and scientific approaches and concepts for modernizing the Philippine system. Increasing pressures in the late '60s and early '70s for speeding up the modernization process led to efforts to introduce more "dynamic" policies, that is, policies aimed at accelerating the development process by enhancing its comprehensiveness, intensity and efficiency. This thrust in policy was made possible because of the growing entrenchment of technological and scientific approaches among political leaders and because of a supportive educational system. The technological and science-based outlook was introduced by a group of "technocrats" who were technically trained experts and professionals, particularly in economics, and who were located in decision-making positions. Through them technocratic ideas began to permeate the entire government system. At the same time ideas of development were turning in a technocratic direction.

The methodological tools used in this work are based on the usual methods of social science. They include the use of descriptive, analytic and historical approaches. The earlier chapters are essentially descriptive. In the selection of the case studies certain criteria were followed, namely, that the cases should be identified with a dynamic thrust in development, and that as much as possible they should cover the experience of ordinary people to allow generalizations about how people see themselves affected by technocratic thought and activities. This work draws on a variety of sources. The principal of these are: literature in the field, the statements of policy-makers and leading administrators, official documents and reports, unpublished documents and interviews.

The Philippines has been selected for this work not merely because it is a developing country but because it has a long tradition of both modernization and technocracy. It was perhaps the first country in Asia to embrace thc concept of modernization and to make an explicit commitment to it. It also produced an aggressive group of young technocrats at a relatively early date, in the 1960s, prepared to experiment with new, dynamic, and flexible ideas of development. Another reason for choosing the Philippines is that the country is a gold mine for students doing research in this area. Records indicate a strong government commitment to the rhetoric of modernization and technocratic values from roughly the late 1960s and the presence of a vast number of projects and programs undertaken by technocratic planners under the umbrella of some kind of development policy, which are open for inspection to students in this field and could, in many cases, make an impressive contribution to the "development debate." Indeed, it may be difficult to find government projects or programs which have not been classified as developmental and so, presumably, outside the orbit of technocratic interest.

The argument of this work is developed in eight chapters. Chapter II reviews the early "liberal" phase of Philippine development thought and its relatively static nature in the context of liberal politics and values. It also describes the "time of turbulence," a period of unrest, the emergence of technocratic planners, and the emergent interest in science and development administration. Chapter III illustrates the new government dynamism in politics, administration and economic and social policies during Marcos's authoritarian regime. Chapter IV presents a number of short studies to show the impact of the new policy dynamism on existing development ideas and values. Chapter V identifies the main causative factors of technocratic thought. Chapter VI considers the extent of technocratic influence in policy-making and Marcos's position with respect to national technocrats. Chapter VII reviews the nature and weaknesses of the technocratic mode of thought and identifies the implication of technocracy for human freedom and values. It also explores the possibility of overcoming the " gap" in values between technocracy and " humanist" ideas of development and suggests a new conceptual approach that might help to reconcile the two values. The last chapter briefly covers the Corazon Aquino era and brings the story up to the early 1990s.

CHAPTER II

Perspectives in Development Thought: the Liberal Phase

Development thought in the Philippines has undergone periodic, albeit gradual, changes in its perspective. This chapter concentrates on its perspectives in the 1950s and 1960s, referred to here as the "liberal development phase" to suggest the general framework of "liberal values" within which ideas about development manifested themselves. It presents a survey of development ideas of Philippine presidents during this period and highlights changes in conventional ideas during the "time of turbulence" of the late 1960s; it also describes the rise of interest in technology and science, the emergence of technocratic thought and the establishment of development administration.

Philippine development thought during this period was shaped by the particular postwar social and political settlement favoring the prewar landowning interests. These were again in power and enjoyed the support of the rising industrial interests. Thus the system was dominated by the agricultural aristocracy and the emergent industrial class, representing an "oligarchic" rule. This explains the inability or unwillingness of the regimes in power to deal effectively with issues of peasant unrest due to the breakdown of patrimonialism in the rural areas which eventually escalated into the "Huk rebellion" in Central Luzon. To this internal settlement the close external relationship with the United States can be added, which led the country to increased dependence on the former colonial power, both economically and politically, in a relationship which some writers describe as "neo-colonial."[1]

The postwar arrangement also included a commitment to institutional democracy. Democratic values commanded universal consent, although they were probably understood in different ways by different classes. The dominant classes conceived institutional democracy in terms of an essentially liberal political system, and an area where political action took place and power games were practised. For them it was, in its existing form, the optimal political system. This belief was supported by the idea that the complexities of an industrial economy required pluralist democracy. But further political development was effectively arrested by the social and political settlement mentioned above, which favored the forces of restoration and excluded the Huks and the progressive interests.

Some changes in overall political values did occur, however, under the influence of an expanded vocabulary of world politics; this, as Lucian Pye put it, "has been engulfed by terms and programs related to a new view of nation building—the developing areas, the emerging nations, Afro-Asian states, economic development plans, charismatic leaders, technical assistance."[2] It included the concept of the nation state, which was now treated as a central political concept. Similar ideas developed in the Philippines, and led to attempts by national leaders to strengthen the standing of the country both internationally and domestically; in the latter case they sought to expand their political appeal and legitimacy by popular democratic means in addition to traditional claims based on patron-client authority. In addition, the power of the state was strengthened. The state became more active in policy-making and in developing activities, which were to lead to a considerable change in its meaning.

Commitments to Development

The first period following the establishment of the Philippine Republic in 1946 until the early 1970s reveals increasing concern with issues of national development and converging views on the objectives of development.

This will be illustrated in the following discussion, which surveys the major policy commitments of Philippine presidents during this period. These commitments largely reflect the political rhetoric of the time.

The development ideas of Manuel Roxas and of Elpidio Quirino, the first two presidents of the Philippine Republic, were shaped by the postwar problems resulting from the country's liberation from Japanese occupation, the dislocations in the economy caused by the war, the acquisition of political independence and later the challenge to government authority posed by the Huk movement. This situation dictated a focus in politics on matters of immediate concern, for the viability of the new nation had still to be proved. These ideas therefore centered on the basic issues of political and economic stability.

Stability, expressed by appeals to "rehabilitation" and "reconstruction," was the main message of Roxas's inaugural address of 1946.[3] In this document three out of eight principal policy recommendations concentrated on external recognition and internal security and included calls for closer ties with the United States, adherence to the United Nations Organization, and the maintenance of peace and order. At least three policy objectives identified general goals of development, which, with modifications, became a recurring theme in virtually all subsequent programs of Philippine presidents. These were: relief for the masses, social justice for the working class, and honesty and efficiency in the government service. The next objective aimed at "reconstruction of the war-devastated economy"; this reflected an immediate concern for the survival of the new state as a viable political unit and its long-term prosperity. The last objective aimed at upholding the country's commitment to democracy. Roxas described this as a commitment to the "preservation of the individual rights and liberties of the citizenry" and expressed it with great rhetoric and passion, as the following passage indicates: "With the help of God, let us build this our land a monument to freedom and to justice, a beacon to all mankind." This passage seems not only to have expressed his personal belief in the superiority of democracy but also to have reflected the pulse of the

Filipino people who had just won the right to democratic self-existence, although, as Roxas's critics perceived it, his "democratic" rhetoric masked the re-establishment of an oligarchic rule.

Specifically, President Roxas affected the future course of Philippine development mainly in the area of the economy. Under his leadership the Philippine Congress was asked to change the Constitution to accommodate American economic interests (which was made a condition for obtaining a large American "rehabilitation loan"). These interests were specified in the provisions of the Bell Trade Act passed by the U.S. Congress in 1945. The main provisions were the so-called parity rights, which granted the same rights of access to American citizens and corporations as to Filipinos in the exploitation of Philippine natural resources. Among other provisions were rights to free trade between the two countries and a fixed rate of exchange at two pesos to one dollar which could not be changed without U.S. approval. These proposals were bitterly decried by "nationalist" politicians and progressive parties as offensive to national sovereignty, but their opposition was repressed by Roxas (some members of Congress were expelled). The necessary change in the Constitution was subsequently approved in a plebiscite. During Roxas's presidency the oppression of the Huks and of progressive parties was a deliberate government policy and led to civil war in Central Luzon. In sum, Roxas's thought suggests the centrality of two ideas, liberal democracy and economic stability, but these ideas must be conceived in a political context that reflects not a restructuring of the old social and political order but rather its revival and continuation in another form.

Stability was also central in the development thought of President Quirino. Its achievement had largely eluded President Roxas, who died suddenly within two years of taking office. The stability theme is present in all the major programs and policy commitments during the Quirino Administration.[4] In this Quirino followed his predecessor, highlighting two general goals, economic restoration of the nation and political stability. The continued preoccupation with political stability was necessitated by the rising "rebellion" of the Huks in Central Luzon. Among his political objectives were: the restoration of peace and order, the preservation of national integrity, the eradication of graft and corruption from the government service, and "honest enforcement" of tax laws. Similarly the preoccupation with economic restoration was necessitated because of a crisis in the Philippine economy that occurred in the late 1940s due to the effects of the free trade policy. To counter this crisis Quirino introduced foreign exchange and import controls, which the Americans allowed as a "temporary" measure. A more lasting economic stability was to be effected by reforms in the economy. For this purpose an American team of advisers, known as the Bell Mission, was asked to provide a groundwork for the desired reconstruction. The mission identified the principal weaknesses in the Philippine economy and recommended a number of important economic and administrative reforms,[5] which Quirino then turned into a set of national policies. These included proposals for encouraging investment in productive enterprise, for coordinating the fiscal, credit and investment policies to prevent the recurrence of inflation, for administrative and taxation reforms as well as a host of other similar reforms intended to increase the efficiency of the economy and government administration. Most important, a substantial new aid package was made conditional on the acceptance of American supervision as to how the aid was to be spent.

Quirino's other well-known reforms included the establishment of the Central Bank during the first year of his presidency (in 1948), which centralized all national monetary activities, and, in public administration, an initiative to reorganize government administration at all levels and establish new government organizations to deal with problems arising in the context of development.[6] Another action that was likely to affect the course of Philippine development was the decision to strengthen further the existing intitutional links with the Americans. In addition to the strong economic links favoring American interests mentioned above, new military links were established through U.S. military assistance, and the

Central Intelligence Agency (CIA) was allowed to assume an open role in matters concerning the country. In his overall outlook Quirino did not depart substantially from the outlook of President Roxas. Unfortunately, his promise to bring about an "era of progress and plenty" could hardly have been redeemed, given the political and economic instability during his presidency.

The same general approach, centered on national stability, was followed by Ramon Magsaysay, Quirino's successor. This is reflected in his ten-point program or "guide-posts," which, according to him, were meant "to stabilize Filipino society."[7] Magsaysay expanded the range of policy interest to new development programs and strategies. They included specific government programs for improving educational facilities, increasing democratic participation, accelerating land reform, alleviating poverty and enhancing the efficiency of the economy and government systems. They highlighted the drive for self-sufficiency in primary foodstuffs, for reforms in the education system "geared to the requirements of scientific and economic progress and the need for scientific research" as well as an urgent need for fiscal and economic development. With variations these ideas can be found in all early proposals for government reforms.

Among Magsaysay's best known contributions to development were his reforms for depressed rural areas and his proposals for agrarian reform, which, according to him were aimed at "bringing freedom and progress to our barrios." His rural reforms included basically three aspects: a land reform and resettlement program; provision of easy credit to the peasants, building country roads and providing technical advice to farmers; and a community development program based on self-help. A major feature of Magsaysay's land reform was a provision encouraging tenants to shift from tenancy to leasehold so they would now pay a fixed rent much lower than the old obligation. Also, tenants could not be ejected except for "just" cause to be determined by the Court of Agrarian Relations. But despite some initial success, his land reform scheme ultimately failed because the basic issue of tenancy was never resolved. The legislation was riddled with loopholes which made it ineffective.

In his economic and political outlook Magsaysay reflected conventional ideas about development, closely following his American advisers and friends. Economically, this is reflected in the Laurel-Langley Agreement signed in 1954, which extended the parity arrangement to all American business activities in the Philippines. Politically, one illustration is Magsaysay's Vietnam policy. He aligned himself with the United States in favor of intervention despite considerable opposition at home, following the American position that this issue was one of " free nations fighting against communist aggression in Southeast Asia." Another illustration is the close personal relationship between Magsaysay and the Americans, particularly the CIA's Lt. Col. Edward Lansdale, in a number of schemes, including American support for Magsaysay's candidacy for the presidency, the suppression of the Huk rebellion (" successfully" ended under Magsaysay), for a program of cooption to complement the counter-insurgency campaign (organized by Landsale), Magsaysay's rural community development program where the funds for all related projects came from the U.S. Aid Mission, the CIA[8], and so on.

Recognition is usually given to Magsaysay's contribution to Philippine political development. With his dynamic and charismatic leadership and image as a man of the people, he enjoyed enormous popularity which was strengthened by well-orchestrated publicity. This led to his great electoral victory of 1953, in which he captured the allegiance of the " little man" of the Philippine countryside. Such cross-national electoral appeal is sometimes viewed as being contrary to the traditional pattern of Philippine politics based on loyalties to local dynasties and as the starting point of the " democratization process of Filipino politics." It demonstrated that the hold of local political bosses over the people was not inevitable, and it helped to turn Philippine politics in a more national direction. Such national orientation was, of course, not new. It was one aspect of the ongoing process of political development caused by the impact of national institutional democracy with the paraphernalia of election campaigns and national symbols; to some extent this eroded the automony of local patron-client links and led to the creation of a national system of power-brokering among regional rather than local warlords. Through Magsaysay's popularity the new relationship became more personalized. During the Magsaysay era political development was also to be advanced by increasing the decision-making powers of rural barrios (villages), which aimed at strengthening "democracy at the grassroots."

The development thought of Carlos Garcia, Magsaysay's successor as president, is reflected in the "keynote" of his governmental program.[9] This was focused more on economic and fiscal policy matters than on broader issues of development. Garcia's preoccupation with fiscal and economic issues must be seen in the context of serious economic and fiscal problems facing the country during his administration. These included a steady erosion of the system of exchange and import controls (caused partly by easy credit policies and liberal exchange allocations prior to the presidential election in November 1957, by which Garcia sought to ensure electoral votes), a severe balance-of-payment crisis in the winter of 1957–1958 and an increasing depreciation in the free market rate of the peso.[10] Thus economic and administrative stability again became the primary goal of Philippine development. Garcia's policy prescriptions were conventional and included three objectives: austerity, moral rectitude and professional competence. By "austerity," he meant "more work, more thrift, more productive investment and more efficiency . . . less imports and less extravagant consumption."[11] His call for "moral rectitude" referred to efforts "to minimize and, if possible, to eradicate graft and corruption"[12] in government administration, while his emphasis on professionalism was based on the belief that success in development called for the upgrading of professional competence. This, according to him, would provide the necessary "assistance in planning and carrying out any comprehensive program of economic and social development."[13] The reference to the need for moral rectitude reflected the rising problem of "graft and corruption" during Garcia's administration. Indeed, this became one of the big political issues, and was a factor in Garcia's defeat in his second presidential election attempt.

An original note in Garcia's administration was the much heralded "Filipino First" policy, which was intended to stimulate domestic productive activity and encourage people to buy Filipino-made goods rather than imported products. This marked the first attempt at presidential confirmation of economic nationalism and led to the industrial policy of import substitution. It was hailed by Filipino nationalists and condemned by the advocates of free trade and the export bloc in Congress. However, Garcia soon had to tone down the nationalist language of this policy because of powerful pressures by American and domestic special interests. Indeed, threatened by reprisals, he had to agree in 1959 to gradual import decontrol, which was later followed by the devaluation of the peso.

Garcia's concern with broader issues of development was institutionalized in the *Five-Year Fiscal Plan 1957–1961*. Although the plan was heralded as an instrument for "maximizing and accelerating the nation's social and economic development" and at aiming to banish unemployment, poverty, disease, and illiteracy from our land,[14] it was focused not on social welfare but rather on measures for enhancing the country's economic growth. It is evident that the thinking underlying the plan followed the tenets of the dominant economic theory in which improvements in living conditions were seen as dependent on increased economic productivity rather than on direct government intervention. At the heart of the plan were "six fundamental fiscal policies and principles" which were offered as "controlling guides" for development planning. Crucial to the success of the policies was the ability to generate a sufficient rate of economic growth. Ultimately, the plan's objective was to provide the right climate for capitalist growth; in the statement of an official spokesman, "the vast machinery of the government [has been geared] towards intensive national uplift by improving and extending its wide range of public service, by increasing the output of public undertakings, and, above all, by creating the right social, economic, and financial climate in which business may venture and flourish."[15]

Macapagal and Early Marcos

Reversing the Filipino First policy of President Garcia, Diosdado Macapagal reverted to a more "traditional" or pro-American economic and political style of Philippine leadership. His development thought covered a wide range of interests and activities and injected a new dynamism into Philippine politics. His public commitments show that he balanced conventional commitments to economic and fiscal efficiency with populist ideas of the "little man" going back to the Magsaysay era. He revived calls for moral regeneration in the public service and extended greatly government commitments to social and economic development.

Macapagal's approach to national development was embodied in his *Five-Year Socio-Economic Integrated Program*, the principal theme of which was the improvement of living standards and future-directed growth, as well as in his numerous public addresses with catching titles, which usually struck the correct political note by identifying aspects that were subsequently to become significant trends.[16] The program contained a conservative emphasis on economic and fiscal stability and gave a substantial place to policies aimed at socio-economic and socio-political development. The desired stability was to be achieved mainly through fiscal policies and free trade. It included economic decontrol, devaluation of the peso and the abandonment of import substitution for an export-led program with increased American participation. The socio-economic proposals concentrated on a number of policy objectives: higher incomes, more adequate housing, particularly for the lower-income groups, production of low-priced clothing, stabilization of prices, expansion of education and public health, self-sufficiency in rice production, encouragement of employment and establishment of cottage industries. One important proposal, following Magsaysay's earlier scheme, was an ambitious land reform program, passed in 1965 as the Land Reform Code; this aimed at abolishing the share-cropping and tenancy system of agriculture and creating instead leasehold farmers of "family-size" farming units as well as providing the necessary technical assistance and credit facilities. Land would now be leased and farmers would pay 25 percent of the harvest to their former landlords. This scheme was first tried in selected areas in Central Luzon and initial results were encouraging. However, like the earlier scheme, this one was riddled with legal loopholes and inadequate funds impeded its implementation.

The socio-political policy proposals of this program were focused on the Philippine village. One of these was to encourage community development, reminiscent again of an earlier Magsaysay scheme, where improvements in rural areas were to be accomplished by "self-help." Its aim was "to develop an alert, self-reliant and resourceful citizenry, capable of improving themselves and their community with a mimimum of government assistance."[17] This was accompanied by a political scheme, again following Magsaysay, aimed at increasing the power of local authorities, particularly at the barrio level. Its stated goal was to transform rural barrios into self-reliant communities with the capacity to govern themselves. It declared that people should be able to determine their own needs and problems, plan their own solution to them, utilize locally available resources such as human labor, new materials and cash, and also to do the work themselves.[18]

Another feature of Macapagal's program was the scheme of "moral regeneration" or "moral renaissance." This sought to address the issue of corruption in government and was part of an effort "to correct the decadent state of our public morality." It aimed at making public officials more responsible to their clients, more honest and upright in their conduct and more efficient in their work. The scheme was accompanied by administrative reforms which aimed to provide "an effective machinery for systematic improvement of administrative organization and management in the government."[19] Through this scheme Macapagal expressed with new vigor essentially conventional ideas for improving government administration, based on appeals to morality and efficiency, and made such ideas part of wider efforts at institutionalizing administrative reforms.

Viewed as a whole, Macapagal's policies did not depart greatly from the conventional framework of social and political thought. While they revealed fresh sensitivity to issues of social and human interest and used language usually associated with later concepts of development—for instance in his call for " massive productive and labor-intensive projects"—Macapagal's policies were nonetheless conceived along conventional lines. Basic social and political arrangements were taken for granted and remained intact. Most of his innovations, such as they were, could be explained as pragmatic. Macapagal's scheme of self-reliant rural communities can perhaps best be viewed as a shrewd attempt to keep the peasant population working and busy rather than as an attempt to create new structures for bringing about greater social justice and equality. Similarly the emphasis on land reform may be considered a pragmatic response to the grave problems of rural poverty rather than a manifestation of rising social consciousness. Macapagal's message of moral rejuvenation followed a well-established public use of symbols to highlight desirable administrative conduct.

In Macapagal's development scheme the role of private enterprise was to be the principal instrument of accelerated development while the government was to play a supportive, although a very important role. He stated that " in essence the program [for national development] will call for a return to free and private enterprise."[20] Armand Fabella, one of the architects of the " New Era" on the administrative side, expressed the same idea: " The basic philosophy underlying the Socio-Economic Program . . . is the primary reliance on private enterprise as the main engine of development, with the government attempting to provide the institutional and administrative framework that would optimize the capability of the private sector in its investment and productive activities."[21]

The early development thought of Ferdinand E. Marcos after his election as Philippine president in 1965 does not indicate any major departure from the established framework. He was a product of the established social and political system and like his predecessors did not question the correctness of prevalent development objectives or the path to achieve them; he merely tried to improve on them. The conventional framework of his early thought is reflected in his inaugural presidential address in 1965 entitled "Mandate for Greatness." Despite a strong call for change, the address did not herald the advent of a new era. It struck a strong heroic and inspirational note through its appeal for renewal of "the vision for greatness for our country" and for a popular "mandate not merely for change . . . [but] for greatness,"[22] intended to impress on the audience the redemptive force of Marcos's new style of political leadership and to symbolize a renewed "moral" and "historical" commitment to the nation's past. However, the rhetoric of this address was ritualistic, a familiar feature of Philippine politics for such occasions; it was reminiscent of Macapagal's references to the "unfinished revolution."

The other parts of the address were also basically conventional. These included recommendations to cultivate frugality and good conduct and practise "self-sacrifice," opposition to "every form of waste . . . of conspicuous consumption and extravagance . . . as inimical to public welfare,"[23] determination to pursue "a policy of rigorous fiscal restraint," and appeals to public officials to set a good example by honest personal conduct. The address made only a brief reference to development and even that lacked clear inspiration and focus. It stated that "economic planning is as essential for freedoms as political planning" and that "we had squandered the energies and resourcefulness of our people."[24]

A more direct concern with development issues is contained in Marcos's State of the Nation address of June 1966 on that theme, but departed little from the prevalent rhetoric of development, except perhaps for the highlighting of organizational and administrative factors in public policy. It showed interest in community development, but the commitment to this was expressed in flat rhetoric. Community development, the document stated, " shall be pursued with greater momentum and depth to develop a new sense of values and to further strengthen our social, economic and political base in an atmosphere of participation and mass involvement."[25] The issue of popular participation was not elaborated. On the other hand, practical concerns with appropriate strategies for community development were given considerable attention. These included government price support for rice and corn, the borrowing of funds abroad to finance the building of irrigation systems and reorganization of government agricultural agencies to reduce their operational cost and attain desired efficiency and " maximum coordination in all levels of governmental planning and implementation."[26]

The other policy objectives addressed here were treated in a similar way. They included investment incentives, better educational facilities, increased national security and increased infrastructure (a concept which later came to be closely associated with Marcos's developmentalist-technological approach). However, they were not clearly related to development planning. The commitments made in the area of government administration were more dynamic. Progressive trends initiated during the Macapagal administration were to be continued and strengthened. While the address made conventional references to the promotion of " economy, simplicity and efficiency in the conduct of public affairs," it also suggested the need of administrative reforms geared " to the requirements of national growth."[27] Like Macapagal, Marcos recommended a reorganization of the executive department.

In sum, development thought during Marcos's early presidency was rooted in the democratic tradition and conformed to existing social and economic arrangements and values. It was expressed in traditional rather than innovative terms. Central was the notion of political development. This was perhaps not surprising. For one thing, it was identified with the robust politics of this era. The mid-1960s were still the golden age of "Filipino liberalism" and major defects in this system were not fully articulated until the late '60s. For another thing, it was identified with democracy, which was believed to be a prerequisite to rapid economic development. Indeed, democracy and development were viewed as complementary, the former stimulating the latter, and both were believed to lead to national progress. At this stage Marcos viewed democracy in essentially individual, liberal terms. In a public address in 1966[28], for example, he identified the democratic idea with the dignity of man, as a "truly radical idea" and as involving a "true revolution" but at the same time he contended that the Philippines had proven that democracy could exist even under conditions of underdevelopment; "democratic institutions," he said, "need not be incompatible with the conditions of underdevelopment—mass poverty, ignorance, and disease."[29] In Marcos's later rhetoric, democracy and development were linked and he asserted the revolutionary potential of democracy for transforming Philippine society in a more egalitarian direction.

Other perspectives in development thought during this period came from such prominent intellectuals as Claro M. Recto and from various more radical political actors and theorists. These people played an important role in the formation of development thought in the Philippines. Recto's political ideas, for instance, highlighted the prerogatives of national sovereignty and national interests as against foreign interests and were directly influential in the formation of President Garcia's policies, while the contribution of radical critics, particularly the Marxists, was through their analysis of the exploitative nature of the existing social and economic order and their proposals for change. However, these actors will not be considered in detail here because this work seeks mainly to define the ideological position of the Philippine government system. Popular ideas on development issues will be briefly identified when necessary for explaining the main argument.

The Early Orientation

The survey of the commitments to development by early Philippine presidents suggests a development orientation that reflected liberal political and economic values and was inspired by Western concepts of modernization. Liberal democracy was worshipped as the best of systems, and economic growth was to be achieved through private enterprise, although the helping hand of the state was always present, and expected to provide a general direction or "protection." The general goal was social and economic stability defined as economic growth and a well-functioning government administrative system. Little change occurred in the rural areas because the impact of agrarian reforms was insignificant, the old feudal order remained dominant, and challenges to this system, such as the Huk rebellion, were repressed. The dynamic aspect of the economy was confined to a relatively small portion of the population. Apart from the traditional agricultural aristocracy, there were emergent commercial and industrial classes and the educated middle classes. The ruling class reflected this social and economic complexity. With the economy expanding, there was considerable optimism in future development, but this was viewed in terms of diffusionist ideas of modernization theory. In this social and political context , the role of government was to be limited. It did not enjoy major planning functions or taxation powers; rather, its importance stemmed from the power to influence the direction of the economy which was exercised through indirect means particularly fiscal policies.

The context described here refers mainly to the earlier liberal phase of Philippine development since the declaration of independence in 1946 till the late 1960s. The later phase, referred to as the "time of turbulence," introduced more assertive and critical ideas of social and political development. Because of its influence on subsequent Philippine development, it will be shown how the early liberal phase manifested itself through ideas of political, administrative, economic and social development.

In politics, development ideas reflected the prevailing "liberal" practice; this was shaped by the demands of practical politics and was essentially a product of particular interests. Jean Grossholtz described this practice as "the subordination of economics to politics," and added that "the planners played little part in development."[30] The politicians' concerns reflected mainly traditional agricultural and the rising industrial interests. The former was represented by the land-based aristocracy—the "caciques"—who were the politically dominant class, the latter by industrial entrepreneurs, some of whom were advocates of economic nationalism, others promoting the priority of export manufacturing. The rise of industrial entrepreneurs and interests had significant political consequences; it gradually undermined the power monopoly of the landowning elite. Golay described this event as "the emergence of a countervailing center which has successfully challenged the agricultural aristocracy on a number of significant economic issues," and added that "the unity of the political elites has been permanently destroyed, and intense competition among the elites portends continued change."[31] In this system, political power was correlated with economic power or wealth. As Golay put it, "Income and wealth are concentrated, political power is highly correlated with wealth, and welfare goals are assigned low priority."[32] Political action was characterized by a lively political style but was confined mainly to dominant traditional and educated middle-class interests. There were multiple sources of political authority and the dynamics of political life assumed the form of a bargaining process. This included a vigorously operating party system and a free press, as well as other paraphernalia of liberal systems. In spite of dissident criticism the system was accepted by most of the population and was viewed by the politically active elites as a dynamic one.

Closely related to these political ideas were ideas about administrative development. This was clearly subordinated to political authority, and the dichotomy of politics/policy-making and administration was recognized. However, in practice this separation was never complete because the system was highly vulnerable to political interference and constantly exposed to the corroding influence of political patronage.[33] This made the system weak and ineffective. An example of this was the use of the bureaucracy to bestow patronage as a reward for loyalty, which led to its expansion beyond the point where it could be adequately budgeted. This in turn encouraged bureaucrats to compensate themselves for inadequate wages, resulting in a vicious cycle of bloated bureaucracy–low wages–corruption. Another example was the practice on the part of politicians at all levels of manipulating the system to suit their interests. Thus the system became "debureaucratized," losing its autonomy, as its goals were subverted by powerful outside interests. Sometimes the bureaucracy was undermined by behavior within itself as agencies deployed political support to increase their resources and influence.[34] In addition, the system was weakened by Filipino cultural and social values, which affected administration at all levels. These tended to perpetuate personalist attitudes and encouraged subordinates to follow personal allegiance rather than functional rationality and impersonality in their conduct. Grossholtz described this as a situation where "power, not function, is the determining criterion . . . and power is determined by connections."[35] Lack of rationality also expressed itself through a strong incidence of nepotism, circumvention of rules, financial corruption and other similar breaches of administrative propriety.

Several attempts were made to introduce reforms into this system. Indeed, this became a recurring theme of Philippine administrative thought, starting with President Roxas in 1947 and repeated by all subsequent Philippine presidents.[36] Actions were also taken to improve management and organization in government systems; in this the Hooverian model of improved O & M (Organization and Management) was followed, including the establishment of central agencies to coordinate government efforts, broaden the scope of planning and evaluate government projects. There were also periodic proposals for streamlining administrative procedures, improving administrative skills and using advanced O & M techniques, particularly in personnel systems. Another initiative was the establishment, with American help, of the Institute of Public Administration under the State University of the Philippines to boost the administrative capacity and standards of government systems,[37] as the crucial importance of administrative factors for the country's development began to be recognized.[38] In the area of administrative conduct efforts were made to reduce or eliminate the frequent breaches of the principle of merit, which could, as one civil service commissioner put it, "nullify the active operation of the merit system,"[39] and to improve conduct by appropriate reforms. These included periodic appeals for moral regeneration in the civil service, embodied in instructions for civil servants to set an example to other Filipinos in honesty, efficiency and simple living.

These administrative reforms were limited in their objective, which was to correct outstanding defects or distortions of administrative systems and immorality in administrators' conduct. Where they went too strongly against dominant political interests, they were rendered either ineffective or were frustrated by the legislature.[40] Attempts at moral regeneration were similarly ineffective. According to Onofre Corpuz, a leading contemporary political scientist, they must be viewed in the context of the traditional way of handling social-conflict situations, as a form of ritualism, aimed at bringing the worst administrative evils within manageable and socially acceptable boundaries. It is also significant that these reforms were "administration-motivated" rather than "development-motivated," that is, they were not yet related to development in the way they became in the 1970s. More often than not, they followed the established pattern of development of administration rather than the new concept of development administration.

During this period ideas of economic development were given prominence and "development" tended to be viewed principally in economic terms. But realistically the economic factor must be seen against the backdrop of a traditional economy. This shows largely a dualistic type of economy with the modern industrial sector concentrated in a few urban enclaves, particularly in the Manila area, and an agrarian sector based on traditional subsistence farming which employed the vast majority of the country's population. Apart from export crops like sugar and coconut the rural economy was relatively stagnant. The traditional peasant-landlord dichotomy persisted as the base of rural social structure, which led Onofre Corpuz to observe in the early 1960s that the country's economy was "the classical economy of developing countries." No special effort appears to have been made to bring the village from its relative isolation into the orbit of the national economy, except for a variety of community development schemes. The official concern for economic development was limited to the small area of industrial policy because of the government's policy favoring industrial development; in Golay's statement, "the goal of national economic development envisaged by the Filipino elite is dominated by industrialization."[41]

Despite the limited scope of government economic policy, the postwar decade witnessed considerable economic expansion. This was stimulated by a great deal of American money coming into the country as war compensation, development aid and investment, and led to the emergence of the industrial entrepreneur and industrial development. The expansion consisted of three periods: economic reconstruction, roughly during the first five postwar years, institutional reforms and economic consolidation, covering the early 1950s, and the subsequent economic expansion.[42] Two events played a particularly crucial role in economic development during this period. One was the promotion of economic nationalism—the Filipino First policy. The second was the introduction of the policy of foreign exchange and import control. Both gave protection to domestic industries and encouraged import substitution, but, combined with exchange parity (of the peso with the US dollar), they made Philippine products internationally uncompetitive and so led to the country's virtual exclusion from international markets.

Within this policy framework, the brunt of economic development was borne by private enterprise which the national leaders periodically declared to be the cornerstone of Philippine economic activity. At the same time, wide responsibility for guiding the economy was left in the hands of the state. Such government "guidance" was not new; it was a part of traditional culture, an outgrowth of governmental paternalism.[43] This now tended to take the form of the state assuming stronger regulatory and controlling powers and the use of new interventionist strategies. "The state," says Golay, ". . . tends to regulate the intensity of the profit motive."[44] Thus, despite its dominant position in economic development, private enterprise was not allowed to operate in an unrestricted way. Indeed, the classical liberal *laissez-faire* policy had never been the operating principle of the economy. Government intervention naturally included a host of regulations of public policy such as the power to tax, spend, manipulate economic incentives, set up state corporations and the like. Still, the scope of such intervention was rather limited; as Golay put it, "the Philippines has had a long tradition of minimum government."[45] Another reason was that, although development planning was introduced as early as 1946[46], attempts at centralized and comprehensive national planning were largely resisted as legislators in Congress opposed any increases in central authority.

During this period, new ideas of social development did not play a major role. The government's interest in new social policies and social justice was only moderate, confined mostly to education and lower-income housing. Issues of social justice were omitted from political discourse and policy action. At the same time there was a growing awareness of human rights based on American liberal values.

* * * * *

Two well-known contemporary works written by prominent social scientists, Onofre D. Corpuz's *The Philippines* and Jean Grossholtz's *Politics in the Philippines*[47] crystallize many leading development ideas of the age and capture with skill and clarity the optimistic intellectual mood of the time. While Corpuz wrote from the perspective of changing political culture, Grossholtz described the existing pattern of political development.

Corpuz's discussion concentrated on the state of development in the early 1960s and the future prospect of development which he defined as "modernization." Corpuz gave recognition to the beneficial effect of modernization in a developing country such as the Philippines, but at the same time he cautiously advised that "our expectations" should not be exaggerated. He reacted strongly against attempts to identify development with the Western path of historical progress. According to him, this distorted reality and assumed a passive view of Filipino culture, encouraging a view of the indigenous culture "as a passive and rather insignificant component, and to assign to the foreign elements the decisive role in the shaping of Filipino culture...[leading] to the conclusion that there is in fact no Filipino culture per se."[48] It also disregarded the tendency among Filipinos, which Corpuz found supported by history, to borrow selectively and reinterpret beliefs. Corpuz sought to draw attention to the ability of Filipinos to adapt modernizing influences to indigenous life.

Corpuz perceived many obstacles to the process of modernization. One was the backwardness of the economy, another the powerful opposition to major reforms, notably to land reform, by vested interests. There were also major weaknesses present in the political system. Although this could be praised for its love of liberty, its free press and its tradition of political dissent, on a deeper level, it was lacking discipline and any profound value commitment. There was too much permissiveness in Philippine politics for too long and too little serious commitment to structured change. Highlighting this point, he wrote:

> The political process is open and is marked by the exuberance of zestful competition by men and groups playing out a bargaining game that is governed in

practice by permissive and fluid rules. The stress is on permissiveness . . . the people are tolerant. There is no historic tradition of intellectual or ideological commitment for Filipino society as a whole.[49]

Corpuz detected the danger that the political system might just "drift along" rather than concentrate its energy on development because of "an enduring motive of pragmatism and compromise in the Filipino approach to the historical problems of the nation."[50] He also criticized the pattern of Filipino political leadership; in his view its base was highly dispersed so that "no one leader can have pre-eminent power status."[51] This weakened national leaders and diminished their power for the effective action needed for dynamic development. A brighter note in Corpuz's argument was his perception of barrio politics in which he saw a trend towards greater assertiveness. Corpuz viewed this as a promising sign of increasing democratization of the country's political life.

Corpuz's perception of the modernization prospects for Philippine society and politics was thus a mixture of optimism and doubt about the prevailing pattern of national development. He envisaged the need for a new dynamism in the economy as well as in political development if the momentum of modernization was to be sustained and enlarged. He also speculated what the possible solution of the problem of further modernization could be, unwittingly anticipating later developments in the story of Philippine development:

The real and long-term solution would have to come about with a further development of Filipino society. It would require the further advance in industralization, and efficiency in govern-mental administration, all of which would expand further the scope of rationality, efficiency, and impersonality in social relations, as well as diminish the sphere of play of kinship considerations. Alternatively, a short-run solution could come about if a crisis were to break forth, giving rise to a leader who could declare a state of emergency and impose a sense of discipline into his party colleagues, the governmental bureaucracy, and the public at large. In the 1960's [he adds] the long-term solution was not in sight, and the danger of the short-run solution did not exist.[52]

While Corpuz's study of Philippine development was colored by his interest in Philippine society and culture, Grossholtz's study concentrated on Philippine political development. She presented it in rosy colors. In her view, the contemporary political process revealed considerable progress in political development and showed an ability to act as an effective instrument of national development. She attributed this to the presence of "the spirit of bargaining," which in her view was the cornerstone of the Philippine political system. In her words, "politics is bargaining and the development of the skills required for bargaining has been the key to political development."[53] Her "bargaining concept" was derived from the American political experience; it was based on the perception of a multiplicity of groups or interests, systematically and self-consciously pursuing their interest in a competitive context, using methods of negotiations and ultimately seeking to reconcile their interests. However, the equilibrium that is attained can be only temporary as the power position of such groups or interests changes. Grossholtz believed that this method of problem solving injected dynamism into the Philippine political and social system. She did not deny that this dynamic competive political game was conducted mainly within the dominant elite; the reason for this was that in the present state of transition the practice was still short of the ideal conditions of political modernization; as she put it, "the requirements of traditional society . . . are still inadequately correlated with the formal structures."[54] However the prevalent pattern of traditional relationships of the paternalistic kind, the patron-client relationships, static by her standard, was changing in the light of the new trend, often becoming "highly politicized." Grossholtz claimed that this led to "overwhelming reliance on politics to solve problems."[55] She also contended that a closer analysis of the political system "reveals that the popular demands do reach the inner circles of decision-making and do influence public policy."[56] Grossholtz of course welcomed such developments as a sign of a start of a major change toward social and economic modernization.

The implication in Grossholtz's thesis was that political rather than economic factors are crucial to Philippine modernization. She said, "the political process has emerged triumphant over the economic conflicts" and "inequality is no longer an issue."[57] She explicitly rejected the older development thesis, which she identified with Ralston Hayden (1942) and Frank Golay (1964). This viewed socio-economic factors, not politics, as central to Philippine modernization and implied doubts whether it was possible to achieve democracy in the Philippines under the conditions of prevailing social and economic inequalities and social and psychological forces. She was optimistic that the ongoing dynamic process of Philippine politics would help to quicken the overall process of Philippine modernization. In effect the country's social and economic problems were to be resolved through mere operation of its political system.

Grossholtz identified three implications of her thesis that democratic politics (in her sense: political liberalism) was central for Philippine development. The first is that democratic politics greatly strengthens national unity and democracy: "The Filipinos have achieved a degree of national unity and a commitment to democracy that stands in sharp contrast to those of other developing areas."[58] The second is that democratic politics has the capacity to accelerate economic growth. She suggested that this put in question the claims frequently made that only planners and state control of the economy could do this, and added that this "has revolutionary implications not only for the Philippines but for other emerging systems as well. It demonstrates that an energetic commitment to economic modernization is possible through the very methods that economic planners and administrators often condemn."[59] One may recall her earlier statement that planners play little part in Philippine development. The third implication is that democratic politics has the capacity to bring about meaningful social development.

Grossholtz argued that this had effectively led to a peaceful social revolution, for it gave rise to new social classes or interests and so transformed Filipino society and broke the hold of traditional social structures:

> The industrial elite challenged the agricultural export industries' control of the political process and lent financial support to reform movements that would enhance their own interests. The disintegration of the old social-economic-political elite, whose position had already been weakened by the wartime experience, was speeded.[60]

Corpuz's and Grossholtz's ideas of Philippine development in the 1960s provide a useful contrast with ideas of development a decade later. In the 1970s the intellectually ever-alert Corpuz developed (with Marcos) a new ideological position defining a model of the "New Society," of which only some elements were present in his earlier writings, and came to view the liberal past with highly critical, even disapproving eyes.

It is interesting that almost all the features identified by Corpuz and Grossholtz as characterizing Philippine development in the early 1960s had changed greatly within less than a decade and had to be vastly modified in the light of the social and political events of the 1970s. A few illustrations may be given of this change. One is Corpuz's objection in the 1960s to the imitation of Western development models. This was still raised a decade later, but the focus had shifted. The later argument was less concerned with cultural identity, and more with issues of people's participation, social justice and perhaps exploitation of the "masses" by the capitalist system. With the advent of the Marcos authoritarian regime, political bargaining lost its former importance and comprehensive planning became a central feature of national policy-making. However, some of Corpuz's original conditions for further modernization were fulfilled through greater rationalization of government systems and the rise of "a leader who could declare a state of emergency and impose a sense of discipline," envisaged by Corpuz in the early '60s.

Similarly, Grossholtz's apotheosis of Filipino liberalism and her analysis of the Filipino political system were unlikely to be accepted as relevant to the conditions of the decade of the 1970s. Indeed, her thesis of political supremacy became increasingly vulnerable to criticism even in the late 1960s as the old economic thesis, which she dismissed, was again becoming central to the understanding of Philippine development. Undoubtedly, a decade later, her statement about the relative unimportance of planners and administrators would have been found incomprehensible, if not heretical. Also her analysis of the contemporary political system was narrow and inadequate. She did not make clear that the "system" she admired was composed essentially of politically active elites, a growing class, but still a small portion of the population, and she seems to have been unaware of deepening rifts in the economic elite, which should have made her less confident about future industrial development. Moreover, she failed to appreciate the development in the Philippines of *mass* political processes (resulting from an increasing penetration of national politics into the locales and associated mass mobilization) which, more than bargaining, were to play a crucial role in "modernizing" Philippine politics. The arguments of both Corpuz and Grossholtz make it evident that Philippine development had travelled a long distance since the early 1960s and that major changes had occurred in its perspectives.

The Time of Turbulence

The optimism reflected in early development thought was based on the belief that, with the help of the state, the country would be able to pursue a course of rapid industrial development and that this, combined with liberal politics, would inject new dynamism into the process of Philippine modernization. This process would keep expanding more widely "from the center to the periphery" while the society would be able to retain its basic stability. Indeed such "stability on the basis of constitutional democracy" seems to have been viewed by the Philippine social and political elites to be the cause of ongoing dynamic development and to continue in the future.

In reality, a variety of economic and social events were increasing the insecurity of a large portion of the population. The removal of exchange controls and the devaluation of the peso caused considerable hardship for the population in general and local entrepreneurs in particular. The new self-confidence of the landlords in the wake of the Huks' defeat led to increased oppression of agricultural tenants with the accompanying feeling of desperation, which, as in the past, had led to revolts against the existing social order. In particular, there was a deterioration of law and order which manifested itself through increased lawlessness and public unrest. Popular demonstrations against the government became frequent, a series of bombings created widespread feelings of insecurity (the best known of these was the 1971 bombing in Plaza Miranda in Manila at a Liberal party rally where a large number of people were killed or wounded, including prominent Liberal party leaders), local warlords with their "private armies" were a threat to peace and order in the provinces, the communists were increasing their activities, and other incidents of violent dissent were growing.

This situation, perceived by many as a social and political paralysis, affected the popular perceptions of the existing system. The faith that the system would be able, as Grossholtz had predicted, to carry the burden of social change with relative ease by extensive political bargaining was increasingly undermined. The ascendent view was that problems of Philippine development were essentially social and economic rather than political and that they could not be resolved by liberal politics alone. There was also a rising awareness among Filipinos of the uneven course of national development and of the need for change. This was, then, a period of considerable turbulence; new demands were made upon the system for social and economic change and there was widespread anticipation of such a change.

The opposition to the system expressed itself in a number of ways. One was the emergence of a new intellectual climate, in which Philippine intellectuals and social scientists set out to criticize the prevailing state of Philippine development. The result of their criticism was that the academically dominant image of a beneficially functioning Philippine democracy began to crumble. A good example of this is the criticism by Raul de Guzman, a prominent student of Philippine politics and public administration, who wrote in 1971: "Recent political development . . . seems to have shattered the image of the Philippines as a 'show-window of democracy'."[61] According to him, in the 1960s "the Philippines offered satisfying evidence of political maturity and stability along the basis of constitutional democracy . . . through civilian politics and civilian government"[62]; however, in the late 1960s conditions changed dramatically. De Guzman discovered a sense of national crisis widely felt in the country, particularly in Manila and among politically active citizens, which he referred to as "the politics of unrest." He attributed the crisis to a variety of social and economic evils: "widespread poverty, rising prices, unemployment and underdevelopment, conspicuous consumption of the rich, graft and corruption in the government, increasing violence, crime and lawlessness, compartmentalized and delayed justice, and the gap between official promise and performance."[63]

Another form of opposition was the emergence of political radicalism in the cities, which now became more visible and vocal. This form of protest, usually accompanied by demonstrations, aimed at highlighting the evils of the prevailing system, especially political corruption, the abuse of power by the governing elites, the inability or unwillingness of the regime in power to take effective action to improve social and economic standards, and the subservience of national leaders to foreign interests. Sometimes these aims were radical. The formation of the revolutionary "Diliman Commune" by students of the State University of the Philippines was one example of radical dissent whose target was the destruction of the Establishment. Sometimes anti-government actions were supported by groups of moderates disenchanted with the system. The violence often exhibited by anti-government demonstrators, as on the "barricades" of Diliman, must be seen against the backdrop of increasingly violent and brutal police actions and subsequent "witch hunts" of dissident elements.

The growing discontent with the general conditions of life did not lead to universal opposition to the regime in power. Rather, it was confined mainly to the politically active minority of the population, particularly in metropolitan Manila, and to Central Luzon, where the Huk rebellion had enjoyed widespread popular support. In particular, it seems to have stemmed from an emerging consciousness of social and political needs that had not been openly expressed before by a portion of the educated middle classes in Manila. For the Philippine middle class, life suddenly appeared more complex and problematic, and they wished to change it. However, such discontent gained broader appeal and sympathy among many other Filipinos because it addressed itself to such deeply felt social and political issues as poverty, land reform and corruption in public places. In this way it seems to have been perceived by the government as posing a considerable threat to its authority.

At least two factors explain the rising criticism of the existing system and the popular call for change. One was the impact of expanded education. The American-type mass education system resulted in a rapid increase in the number of graduates with secondary and tertiary qualifications. Perhaps their numbers gave them further confidence, particularly in Manila, and they started to look at the Philippine system with more critical eyes. As some came to see it, the poverty of most Filipinos was not inevitable or natural but rather a matter of particular social or political arrangements which favored a few. Some of them became disillusioned with the existing system because, despite their college education, they were unable to find a suitable job and ended up in inferior positions.

The second factor was the increased perception of accumulating social and economic problems. These stemmed from a variety of grievances, and having acquired a considerable magnitude began to be perceived by many people as "crises" in the existing system. Most Philippine social scientists, like Corpuz, explained such "crises" by concepts derived from modernization theory. They held that this was a period of transition during which considerable social tensions and conflicts could be expected to emerge.[64] These were then traced to factors which, the theory says, tend to accompany the modernizing process: growing population, increased urbanization and industrialization, greater occupational mobility, migration of rural people to urban areas, and the like. With so many conflicts or problems arising at the same time, however, this explanation failed to clearly indicate where the actual benefits of modernization were to be felt, or indeed how the process of modernization was to remain the dominant influence.

However, such crises can perhaps be better explained in terms of structural problems in Philippine society itself. The examination of Philippine social structures during this period reveals continuous social and economic changes in progress. These were reflected in such different areas as a shift from agricultural production to commercial practices, a drift of rural people to the cities, overcrowding of urban areas due to the rapid population growth, urban unemployment, the frustration at the failure of modernization to meet the expectations of the people partly because of government economic policies favoring other interests, and so on. Another illustration of these changes was the increasing breakdown of the feudal patron-client order in rural areas. This was eroded in part because of the growing population and migration, which made it difficult for the patron to discharge his traditional obligations; moreover, in the cities such relationships became necessarily more impersonal or lost their meaning as the patron was too distant and sometimes could not be identified. The patron-client system was also eroded by the development of mass politics. One consequence of this was the replacement of this protective social order by the much harsher capitalist regime that predominated in Philippine urban areas. The crises in the existing system seem, then, to be related to the ongoing structural changes which gave rise to structural problems. As these problems could not be resolved, they generated instability and feelings of frustration. This suggests an increasing disintegration of the traditional social order and a failure to replace it by a new stable order, which seems to have stemmed from the inability of the existing system to adequately integrate traditional elements into its "modernization" model.

The disarray in the existing social and political system gave rise to extensive criticism in all areas of development activity as well as to new ideas about desirable reforms. In politics, criticism assumed a multifaceted dimension. The prevailing political system was assailed as too conservative and inimical to change, and therefore incapable of positive response to outstanding problems of the nation. Other criticisms highlighted the unrepresentativeness of the system. Politics was assailed as being a monopoly of particular socio-economic groups, motivated by particularist rather than universal interests and values; because of their vested interests, these groups tended to obstruct badly needed changes in the existing system. Also subject to criticism was the existing practice of political leadership and popular participation. The former was said to be undemocratic, based merely on a circulation of elites, with access to it limited to those who agreed to be "coopted" into the system. Popular participation in public decisions was assailed as merely formal; as de Guzman put it, "the theory of popular sovereignty has not been borne out in the actual operations of the political system."[65] It was also pointed out that there was little difference in the philosophy of the two leading political parties and therefore a meaningful political choice was lacking, while radical opposition was effectively suppressed.

Criticism of Philippine politics came also from President Marcos, who represented the political system at the highest level. In 1969, while he extolled the virtues of Philippine democracy on the one hand, Marcos drew attention at the same time to its failings and recognized an urgent need for its purgation.[66] He criticized the system mainly for its inability to improve the living conditions of ordinary people, for its corruption and faulty management procedures in administration ("inefficiency and immorality in government") and for "excessive partisan politics," which he blamed for being the primary factor responsible for the abuse of democratic principles in the country.[67] Ironically, although himself a master-player in the rather dirty power game that characterized Philippine politics and so part of the system, Marcos was distancing himself from the system's defects. He wished to cast himself into the role of a progressive national leader who recognized, like the progressive public opinion then, the need for major reforms in the political system and values.

Criticism was also raised in the area of administrative development, but at the same time new ideas were explored. It is, however, significant that this criticism did not address itself to such conventional administrative evils as bureaucratic inefficiency or unethical conduct but rather to the failure of bureaucracy to become a dynamic system. According to Abelardo Samonte, a leading student of Philippine administration, the existing public administration system was too static.[68] He admitted that some improvements had been made but these were not necessarily motivated by the desire to create innovative values and development-oriented systems; indeed such innovations would have probably occurred at any rate with normal advancements in the administrative sciences and techniques. In Samonte's view, the system failed to respond to new trends in the economy towards accelerated development, and the reason for this was that "Emphasis [in the past] was placed more on the reduction of government expenditures than on the allocation of the existing resources to support new activities and create opportunities for accelerating national development."[69] Samonte of course implied that public administration should become more development oriented, that is, more responsive to issues of national growth.

A major initiative in this area was taken in 1969 with the establishment of the Commission on Reorganization. This mandated the task of comprehensive reorganization of executive departments and government agencies in seventeen areas, including governmental "instrumentalities." Its review took until early 1972 to complete and reflected the new thrust towards more centralized and development-oriented concepts of government systems. As envisaged by reformists like Jose V. Abueva the system was to depart from the traditional emphasis on "economy and efficiency" and a new role was to be given to it in the context of development. Now administrative structures and functions were to be linked with policies for national development. Abueva described this in the following terms:

> The law establishing the Commission on Reorganization is unique in linking the traditional and recurring reorganization objectives of "economy and efficiency" to the broader "national goals for accelerated social and economic development . . ." It took our leaders some twenty years of political independence, several development plans, and three major reorganizations to make explicit the connection between administrative structures and functions on the one hand and national policies for development on the other. Until lately, "efficiency," "economy," "a simplified or streamlined organization" and "effective execution of the laws" seemed worthy ends in themselves.[70]

The reorganization scheme proposed by the commission was a compromise. In typical Philippine political fashion, it made concessions to new demands for increasing administrative efficiency in government systems; at the same time it was not allowed to undermine various vested interests. On the one hand, some government functions were to be integrated; on the other hand, there was to be increased local autonomy and greater dispersal of government authority. Furthermore, public corporations were to be given as much autonomy as possible to prevent their being used as a source of political patronage by the central authorities. Whatever the reaction would have been in Congress, the scheme was opposed by the advocates of "accelerated development." It was criticized for conceding too much to decentralization and for not strengthening the national system enough. It was also criticized for being conceived too narrowly. This was the conclusion arrived at by Romualdo Tadena in his evaluation of this scheme, that "on the whole the GSRC [Government Survey and Reorganization Commission] operated on the concepts developed by the proponents of 'scientific management',"[71] concentrating mainly on the administrative structures and functional relationships among government agencies without tackling the broader issues of development. On this critical view, the original commitment to the broader "national goals for accelerated social and economic development," mentioned by Abueva, failed to be honored.

A similar critical and also constructive approach characterized reactions to prevailing economic policies. It was argued[72] that despite optimistic predictions, these policies were unable to bring about economic progress and stability. They could not prevent recurring fiscal crises causing instability in the economy, and persistent unemployment and underemployment, and could not achieve and maintain a steady rate of economic growth. The leading economic critic and advocate of economic change was Gerardo Sicat, who later became head of NEDA (the National Economic Development Authority), the chief national planning body. His criticism was aimed mainly at two areas of economic activity. The first was what he called "distortions" in industrial development, by which he meant economic malfunctions resulting from excessive emphasis on import substitution, which then was at the core of the government's policy of industrialization.[73] According to Sicat, existing economic policies were too narrow, related mostly to one area, namely import-substituting manufacturing; yet the expansion of this sector was coming to an end as the sector, because of limited domestic markets, had exhausted its original momentum. The second was the "distorted" pattern of government incentives, which he claimed led to contradictory employment policies. Such incentives helped mostly large-size manufacturing industries and so were biased in favor of capital-intensive investment; yet practically all Philippine governments since independence made employment opportunities one of their principal policy objectives. For this reason, Sicat argued, policies should as much as possible utilize " the country's abundant human resources" as one of the bases of incentives in industrialization.[74]

Sicat's proposals were intended to inject new dynamism into what he perceived as increasingly static economic conditions. They were based on the calculation of comparative economic advantages. This formula for economic progress highlighted three factors: making use of opportunities offered by the country's cheap labor; expansion of industrialization for external markets; and overall rationalization of the economy. The last was likely to increase government control over the economy and introduce more extensive national planning. These reforms would ultimately, Sicat believed, lead to a more even distribution of economic growth and to greater equality and stability. Sicat's argument must be seen in the context of the ongoing debate between economic nationalists who sought protection for local industries and the " trans-nationalists" who advocated removal of all barriers to imports and export-oriented industrial activity. Sicat was on the side of the transnationalists. He saw the economic future of the country in the exporting sector where the " comparative advantage" of the country's cheap labor could be exploited.

New approaches were also explored in the area of social development as social welfare and social justice were becoming central items of debate on development issues. Old commitments to social welfare were increasingly perceived as insufficient to bring about meaningful improvement in the lot of ordinary people. Even writers who were not radicals were becoming convinced that in an essentially inegalitarian society like the Philippines, growing inequality in the presence of rising expectations stimulated by modernization might lead to a violent social explosion. The issue of social development had acquired special urgency because of the emergence of movements of political radicalism. One of their attractions was their promise to eradicate vested privileges and radically improve the living conditions of poor Filipinos. They rejected as inadequate most schemes proposed for social amelioration and often preached a revolt of the masses of the Marxian kind, seeking complete destruction of the existing socio-economic system.

This situation stimulated fresh concerns for social needs and social reforms and led pragmatic Philippine politicians to make appropriate gestures of concern. One result was the passing in 1969 of the "Magna Carta of Justice and Economic Freedom," whose objective was to chart a new course in government policy for the social welfare of Filipinos. In the description of its sponsor Speaker J.B. Laurel, Jr., this document was intended to "provide the nucleus of a social and economic philosophy for the nation."[75] The Magna Carta was comprehensive in scope and broadly covered all major social and economic goals. Among its "paramount objectives of economic policy" were industrialization, maximization of employment, promotion of technology and the growth of per capita income, while its other objectives included more equal distribution of income and wealth and the creation of a "large middle-class." Otherwise the Magna Carta took the familiar line of argument against conspicuous consumption and "the ostentation of wealth" (declared to be "an assault on the social conscience and should be avoided by all") and praised "austerity and self-reliance." It also came out on the side of such standard remedies for poverty as family planning; it revived the liberal revulsion against "unwarranted" monopolies, and advocated a mild form of labor participation in the form of profit sharing in private enterprise and the labor's "fair share of economic rewards." But, as Ocampo pointed out, no anti-poverty program was provided.[76]

The late 1960s and early 1970s, as we have seen, was a period of new perceptions and of awareness of the need for reforms. Yet despite the widely recognized defects of the existing political system, the vast majority of Filipinos did not question its legitimacy. Moreover, the dominant elites did not seem to have lost their faith in the ability of the system to improve the existing conditions. This belief was based on two grounds. One was the rather dazzling experience of Filipino elites with liberal democracy in the two-and-a-half postwar decades. To them this allowed freedom of expression, a lot of political maneuvering, was accompanied by a lively political style and encouraged political rhetoric, and so seemed to give wide grounds for dynamic action and for optimism that things were going to change for the better. The second ground for a belief in liberal politics was historical and ideological. It was associated with the Filipino nationalist tradition which was rooted in the revolutionary struggles against Spain in the late 19th century and was a product of political liberalism. As in the past, the present liberalism was to provide leadership for future development as in Macapagal's concept of "unfinished revolution" and Marcos's concept of "democratic revolution." The heroic past was thus used to give sentimental support to the existing liberal system.

The dominant view of the political system was that of the advocates of modernization like Corpuz, who, as noted earlier, wrote about the process of "democratization of Filipino politics." They anticipated that expansion of modernization, including increased industrialization and commerce, growth of population, increased social and occupational mobility and similar developments, would accelerate social differentiation and the formation of new political interests, which would in turn result in more political participation and more democracy.[77] Another view was to seek to cure the defects of the system not by creating a new system but simply by injecting more democracy in it. Robert Stauffer, for instance, advised "politization of erstwhile passive populaces, encouraging competition and conflict, and risking the overturn of existing elites."[78] According to him, the system could lead to more democratic politics by the extension of the competitive political style from the dominant few to the population at large, broadening the base of selection and recruitment of people's representatives, granting more autonomy to legislative bodies as well as by other means enhancing popular participation. A similar reliance on political cures was advocated by de Guzman, although his position was very different from Stauffer's idea of politicization of the masses. His prescription for injecting a new life into the moribund Philippine institutions and to " produce the desirable basic reforms for a viable political system" was based on the idea that " the country's elite is not monolithic."[79] Hence, he argued, it should be possible to combine the interest in reform generated among the socially " concerned and enlightened elite" with pressures generated from below to bring about social changes " resulting from better social organization."[80]

The problem of what to do with a political system that was not responsive enough to demands for change was made more bearable by expectations of impending reforms in the Philippine polity. The Constitutional Convention (" Con-Con") met in 1971 to prepare a new constitution; it was expected to help overcome the main defects in the political system and open up avenues to deal with the main problems facing national development. It was widely viewed as giving an opportunity to promote institutional reforms by peaceful means that would not only affect some parts of the political system, but would also conceivably change Philippine society as a whole.

This role for the convention seems to have been envisaged by Jose Abueva, the convention's secretary, and is reflected in his " urgent" call to Con-Con delegates " to reorganize the socio-political structure underlying elections, policy-making, and public administration."[81] Abueva's recommendations for constitutional reforms included a number of proposals considered progressive such as giving more political influence to young people, urban voters and the middle classes; breaking the monopoly of the two-party system; making the legislature more representative; the promotion of party solidarity; allowing a free party competition; and, optimally, adoption of a parliamentary system of government.[82] Progressives like Abueva wanted open deliberations about the country's institutional reforms and, like Corpuz, wanted to get away from mere " ritualism" or " the escapist and ceremonial aspects," which, according to Corpuz, plague traditional Filipino politics and account for the wide gap between appearance and reality in the Philippine system of government.[83]

Thus in the early 1970s politics was expected to play a key role as an agent of social change. Such changes were to be brought about from above by Con-Con delegates acting on behalf of the nation. Perhaps too much was expected from these delegates. For one thing, the public undoubtedly underestimated the extent of the erosion of confidence in the political system. Indeed a question could be raised whether reform proposals of this political elite group could be trusted to reflect the popular interest. For another thing, the presence of the convention was unable to stop public unrest and the breakdown of law and order. Also the extensive advocacy among some delegates of political and economic nationalism (in the Philippines always tightly linked to anti-Americanism), supported by students, was likely to alarm US political and economic interests and the forces of the Establishment. All these signs of turbulence, to which the alleged threat by communist insurrection was added, gave the opportunity to President Marcos to tighten the reins of government by declaring martial law after first suspending the writ of habeas corpus.

Science and Technology

A significant development during this period was the emergence of technological and scientific influence in public policy. Techniques of technology and science had been used in the preceding decades but now they were increasingly given prominence. This change was a gradual process, accomplished by a series of decisions which had a cumulative effect. Most advances were made after the establishment of Marcos's regime in 1972 but a strong trend in this direction had been present even before. The "scientific approach" was, of course, identified with the application of the theories and methods of science.

There were several reasons for opting for the scientific approach to development. One was the belief in the efficiency of science. Advanced techniques of science were believed to offer an energetic, orderly and controlled type of development based on rational decision-making, which contrasted favorably with "unscientific" approaches of the preceding decades. Another reason was that the "Technological Revolution" was by then an established fact, and appeared to offer the necessary know-how for dealing with problems of social and economic change. The Green Revolution was one outstanding example. The choice of this approach was undoubtedly influenced also by the practical consideration of the availability of advanced scientific models and techniques for decision-making. Of particular usefulness were new models in management and behavioral sciences and in economic theory, which included advanced theories of organization and planning.

This approach was also attractive for political reasons. In an overheated and unstable political environment it seemed to offer a way of avoiding confrontations and yet achieving desired results. It allowed dealing with policy issues in technical rather than political terms. Moreover, the new technocratic expert with his low profile and subservience to the governing authority was not perceived as competing with politicians for influence and power. Similar approaches were adopted in other developing countries in the belief that they would accelerate development, and this approach had strong American support. Indeed, the new technology was imported into the Philippines via American technical assistance in the 1950s in line with a development philosophy which was strongly committed to scientific modelling in the areas of government administration and economic planning.

The Advent of Technocrats

An important motor of the technological-scientific revolution taking place in Philippine development was a group of technocrats who became increasingly active and visible from the late 1950s onward both in the private and public sectors. This study, however, deals only with those who were in government employment. Becoming increasingly prominent in top positions in national administration, they had much in common in their backgrounds.[84] They were products of the American educational system followed in the country's elite schools, and many of them had studied in top American universities, particularly Harvard and the Massachusetts Institute of Technology (M.I.T.). They were young, most of them not older than thirty-five years at the time.[85] In their general outlook, they embraced the tradition of American liberalism (in the disapproving comment of a Philippine Marxist critic, " the most rapacious kind of ideology"[86]) rather than the socialist formula of social progress adopted in many developing states. With their American tutors they shared an admiration for advanced industrial and economic systems and advanced methods of technology and science and, like those tutors, believed in the desirability of meritocracy and in the ethos of professional conduct. At the outset the technocratic position was articulated almost exclusively by Sixto Roxas, but the same ideas appear in the speeches and writings of like-minded administrators and more fully in later technocratic thought, which followed closely Roxas's original position.

The early role of Philippine technocrats was primarily as critics of government systems; only later did they become active reformers. Their criticism was directed mainly at the inefficiency of governmental organizations and what they perceived as the irrationality of the way in which the national economy was run. This criticism was raised as early as 1963 by Sixto Roxas. In one passage he stated harshly and bluntly: " The government is plagued with a sticky administrative machinery of long standing" and argued that this made rational planning of development impossible.[87] Two years later he repeated this complaint with added vigor:

> My basic premise is that it is meaningless, at this stage, to talk of national planning . . . The whole public administrative system, the whole governmental machinery as it stands, is a large, cumbersome and

sticky structure that militates not only against *implementing* a plan, but even against *formulating* a meaningful plan. The "five-year plans" we have had during the whole post-war period were not really plans but merely statements of general aspirations.[88]

Roxas's arguments provide an interesting insight into some basic technocratic ideas. They highlighted particularly the relationship between planning and organizations. Roxas emphasized that planning is necessary for national development. However, taken by itself, planning is not sufficient for successful plan formulation and implementation. What is needed is a vigorous organization, hence his contention that the current "crisis" in development is "a crisis of public administration, of government management."[89] Accordingly Roxas centered his interest on the apparatus of administration and on increased managerial capabilities and efficiency. His proposals for "removing the bottlenecks to development" involved a set of dynamic economic and administrative reforms. These included the creation of an expanded and aggressive capital investment program, more systematic and efficient administration of public services, better administration of the financial assistance and encouragement of the private sector.[90] Typically, he gave little consideration to the political dimension of his proposals.

Early technocratic thought reveals at least four general characteristics. The first is the view that the state should be an active and dynamic participant in the process of development. Although advocating a market economy, Philippine technocrats were not blind believers in unrestrained competition. They tended to view governmental intervention as, on the whole, beneficial for modernizing the economy, provided the state limited such intervention to what it could, in their opinion, do well, namely, establish optimal conditions enabling private enterprises to grow and flourish. (This, however, was later to be massive, including massive investment in the country's infrastructure and the creation of extensive government monopolies and other enterprises.) Hence their advocacy of close cooperation between private enterprise and government planning.

The second characteristic is the advocacy of comprehensive planning conceived in a comprehensive way and intended to bring the entire process of national development under government control. This was expected to eliminate or minimize uncertainty and maximize the rationality of systems, thus resulting in increased productivity.

The third characteristic, closely connected with planning, is the emphasis on management and organization. Indeed, for Roxas and his technocratic followers organization and management were of central importance; they tended to view problems of development in essentially organizational terms as having to do with adequate planning, management and coordination. It is perhaps no exaggeration to say that the technocratic revolution was essentially an organizational revolution. The technocratic notion of organizational rationality went, however, beyond conventional organization theory. Unlike the rigid Weberian hierarchies, technocratic organization structures were conceived as flexible, adaptable and goal-oriented, and their flexibility and adaptability were identified with non-dogmatic, pragmatic approaches, which were believed to be necessary under conditions of rapid change. Organizational rationality also meant commitment to macro-approaches. This went again beyond its conventional meaning, based on a narrow legalistic preoccupation with systemic economy and efficiency. In Romeo B. Ocampo's fitting description, Philippine technocrats were "scornful of the traditional basis of administrative rationality and its distortions—of the myopically legalistic mind, the accountant's 'check-point' or 'GAO [General Auditing Office] mentality' with which older government professionals have handled resources."[91]

Organizational rationality was also equated with the utility of large-size organizational units. This was justified by the economics of scale as maximizing productivity, and followed the popular trend in America towards oligopolistic economic arrangements. Indeed, the organizational focus followed prevailing organization theory in America in which the corporate philosophy was much admired. The figure of the Organization Man became a significant part of the domestic socio-economic folklore, and large-scale corporations tended to acquire new respectability, based on their reputation for dynamic use of resources and high professional competence. Traditional hostility to them seems to have been modified by the doctrine of countervailing power, which suggested the possibility of keeping them under public control (e.g., by "balancing" their monopolistic power by such "countervailing" bodies as big and strong trade unions or large consumers' associations).

The fourth characteristic of early technocratic thought was an avowedly non-political outlook. Technocratic planners seemed to view the undisciplined politics of the time with considerable distrust, even distaste, as the principal villain responsible for the sluggish and uneven results in Philippine development. Indeed, this was perhaps a classical case of a Weberian conflict between administrative rationality and the appeal of politicians to "irrational" and "unpredictable" popular sentiments. In Roxas's phrase, "Within wide limits, there is a basic conflict between the political need for visible display of accomplishment . . . and the requirements of sound long-range planning."[92] Technocratic planners objected mainly to the unscientific character of political decisions in development policies, to the tendency in politics to follow mainly expediential criteria. "Traditionally," says Roxas, "the allocation of governmental resources was innocent of scientific economic considerations . . . of economically rational principles. Expediency and political pressure were the paramount bases for decisions in these matters."[93] Hence, the preference shown by politicians for projects with high political appeal but low economic productivity.

It is significant that despite their distaste for irrational political practice, Philippine technocrats did not openly assail the political system. At least national leaders were prepared to listen to them and to adopt some reforms, and technocrats were increasingly coopted by political elites and themselves became part of the country's Establishment. More importantly, their conduct can be perhaps explained by a basic similarity of interests and values. Both politicians and technocrats shared a belief in the efficiency of private enterprise and the market economy. Moreover, Philippine technocrats were aware of the advantages of political liberalism for economic liberalism. There was, then, no basic conflict of interest between them. Capitalism made more " rational" or dynamic is what the technocrats wanted, together with a sympathetic political system to give them support.

Philippine technocrats were not, of course, uninterested in the country's political life. A leading technocrat-administrator, Armand Fabella, tried to come to terms with problems created by liberal politics on technocratic terms. He made a proposal in 1964 to Philippine politicians, which, in the context of politics then, appears politically unrealistic, if not naive, challenging them to shed their spirit of partisanship in favor of broader vistas of national economic development:

> Another *quasi-economic* objective is a reorientation in internal political thinking. It is important that regardless of which vested group or power group holds the reins of economic power, economic power should be viewed *qua* economics rather than *qua* politics. That is to say, our public officials, in making economic policy-decisions, should endeavor to ignore temporarily the fact that they belong to a particular political party or group. This would make them less motivated to put impediments into the implementation of an economic development program already started by a previous administration controlled by the opposition party.[94]

Fabella sought to draw a sharp distinction between political interest and economic-administrative interest, highlighting the importance of the latter. This led him to make the paradoxical suggestion that it might be desirable to " de-politicize" Philippine politicians, an attractive fantasy for Philippine technocratic planners and administrators. In highlighting the superiority of organizational rationality over other values, Fabella's argument nonetheless does get to the core of technocratic principles for policy action, the view that problems of organization and productivity should supersede political and ideological problems and that political and ideological problems can ultimately resolve themselves into questions of organization and production. As Ocampo fittingly put it, "Questions of traditional political theory are cast in terms of a model that robs them of their political character. Thus, the politics and state of technocrats is ambiguous."[95]

Already, at this stage, the technocrats by their call "to ignore politics" took a position that revealed the limitation and bias of their thought. They had a narrow conception of the functions and values embedded in political systems and were tied to a modernization theory of development, thus never really played the neutral role of "experts" they conceived for themselves. However it should be said that the denial of value to politics did not express their covert support for the status quo against the opposition. Only political systems based on ideas harmonizing with technocratic values could be accepted by them as rational and "desirable politics."

By the start of the 1970s Philippine technocrats had grown greatly in number and influence. A tribute to their success is the directory of technocrats which was published in 1969.[96] The publication contained 526 pages and listed some 3,500 former scholars and grantees under the U.S. Technical Assistance-National Economic Council program including "local scientific, educational engineering and technocratic communities."

Development Administration

Closely connected with the emergent technological-scientific outlook was the emergence of development administration. This innovation in administration theory was inspired by the need to increase the efficiency of administrative systems in the context of development.[97] Variously described as administration for changing societies, "a carrier of innovative values," "an agent of change" or "management of change," the approach gained increasing recognition in administration theory. In Weidner's well-known definition, which suggests its general aim, development administration is "the process of guiding an organization towards the achievement of progressive political, economic, and social objectives that are authoritatively determined in one manner or another."[98] Defined simply, it is the administration for carrying out development objectives—policies, programs and projects—in developing countries, where administrative systems operate under conditions of a wide range of demands for action, but suffer from low capacities and other severe obstacles which are due to their lack of development.[99]

The concept "development administration" may be analyzed into a number of aspects. The first is goal orientation. Administrative action is related to pre-determined goals of development. The second is the emphasis on innovation and change. Administration is related to change in systems; it should facilitate change. This point is summarized by Weidner: "development means change" and "development administration is concerned with maximizing innovation for development . . . it contributes to growth in the direction of modernity or nation building and socio-economic progress."[100] Other aspects defining this concept include the use of rational criteria and standards that are related to development objectives; inculcation of development-oriented conduct and values; organizational flexibility and adaptability to make systems respond to changing conditions; preference for comprehensive macro-approaches, particularly in policy planning; rationalization of systems, and similar organizational and behavioral concepts.

The analysis of "development administration" suggests considerable differences between this model and Max Weber's classical model of bureaucracy as well as its affinity with technocratic approaches. The classical model can be defined in terms of closed organization systems, operating passively and mechnically, and value-neutral in relation to administrative conduct. The new model can be defined in terms of open organization systems, is related to development objectives, considers organizations to be active, responsive and adaptable, and highlights value commitment (to development) in an administrative conduct. Thus the two differ greatly in their general orientation. There is, however, a close relationship between development administration and technocratic models as both advocate advanced economic and management models to increase organizational efficiency in pursuit of development. Development administration adopts technocratic concepts like comprehensive planning and organizational flexibility and adaptability, and tends to view problems in organizational and technical terms. There is also a close relationship between development administration theories and classical economics. The former ally themselves with the latter by linking administrative concepts to economic concepts of output, productivity and efficiency and by adopting notions of rationality from economics. The character of economic thinking in administration is partly due to the great increase in the number of professional economists working in the government service. Through them economic concepts penetrated to the core of the administrative machinery of government, including national planning and decision-making. Indeed, in the theory of development administration, administrative and economic aspects of development form a symbiotic relationship. Only at the later stage did the theories of development administration show concern for humanist ideas of development, but, like technocracy, they did not seem to have been successful in translating such concerns into policy development.

The extent of influence of development administration theory on development activities in the 1960s is hard to estimate. But it was increasing, partly because of the attention paid to development administration in influential professional journals, such as U.N. publications dealing with economic and administrative matters, partly because of increased commitment by national leaders to accelerated development, and partly because of the growing influence of advocates of administrative reforms. A trend in this direction could be discerned in the early 1960s, during the Macapagal era, but only later did it acquire a momentum of its own, reflecting a more assertive outlook.

Among the first advocates of development administration in the Philippines was Rafael M. Salas, a leading official in the civil service. He wrote in 1965:

> The decade of the 1960s may be said to be the decade of the full flowering of the development concept in our country. This concept has been with us in the previous decades but somehow it has been overshadowed by more pressing concerns such as the consolidation of our newly-won independence and the establishing of peace and order throughout the land. Today the development concept has pervaded every area of government activity.[101]

Another early advocate of this movement was Armand Fabella, who came out strongly in favor of professionalism in the civil service, advocated increased economic rationality, including more comprehensive national planning, and, as noted earlier, appealed to Filipino politicians to act impartially.

It is evident that in the '60s and early '70s several developments occurred which deeply affected the course of subsequent national development. They were characterized by the greater influence of science, and of technology and development administration. Ultimately, their application was expected to move the process of development away from its stagnant condition towards a more robust pursuit of national development goals, rectifying the defects in Philippine development identified with the liberal era.

CHAPTER III

The New Dynamism

The declaration of martial law in the Philippines by President Marcos in September 1972 was accompanied by dynamic changes in the political system which influenced development thought in a technocratic direction. This chapter describes the new dynamism as it is reflected in the regime's commitments to development policies and in Marcos's political rhetoric. In the process we will see how this manifested itself in politics and administrative, economic and social development and in the use of concepts and ideas that display the presence of technocratic values.

Strong pressures for political and social change emerged in the Philippines in the late 1960s and early 1970s, but this was difficult to introduce, given the existing institutional framework. This framework however, was about to be modified through the Constitutional Convention which was in session in 1971. President Marcos, not risking the outcome of the Con-Con deliberations, declared a state of emergency in September 1972 and established a martial law administration throughout the Philippines. The principal justifications given by him for this move were that a communist insurgency was threatening to overthrow the government and that there was a general breakdown in law and order.[1] The main immediate reason was to circumvent the provision in the existing constitution prohibiting reelection of a president for a third term (Marcos's second term was to expire shortly). Marcos not only claimed a constitutional mandate for his martial law regime but also presented himself as a leader who came " to save the Republic and reform our society."[2]

The initial steps were taken to stabilize the new regime through suppressing or pre-empting opposition to it. Opposition leaders, Con-Con delegates and journalists known to be anti-Marcos were detained, universities were closed, the old Congress building was padlocked to prevent legislators from meeting there, the press was closed down and a curfew was imposed. Next came the establishment of law and order. This involved a military crackdown on criminals (including the public execution of a Chinese drug dealer), and the disbanding of the "private armies" of local political warlords. Preparations for military action were made against communist "rebels" and other armed dissidents. Some members of the old oligarchy were divested of their property. Such drastic actions were not, however, necessary to defend the new regime. The general reaction of the population was one of quiet acquiescence and "the massive demonstrations against martial law declaration expected by many opposition figures failed to materialize."[3] Indeed many people welcomed Marcos's promises of an end to the perennial stalemate between wrangling politicians, with all the instability and paralysis which this entailed. At the same time the opposition of Philipine nationalists was tempered by Marcos's skillful appeal to a new type of Filipino nationalism represented by the "New Society." Most Filipinos were prepared to concede to the regime the necessary legitimacy, at least during the initial years of its economic success.

Changing Concepts of Development

With the advent of the martial law regime, development concepts were modified to reflect the new dynamic thrust in government policy. "Development" was now conceived as "accelerated growth" and its "comprehensiveness" was given emphasis in national planning. Government programs were expected to go beyond the former liberal, rather limited, preoccupation with social and economic problems and were to be related to national development objectives. In this way development became the supreme good. Indeed, the concept was adopted as a useful jargon; virtually all government activities, including government programs and projects, were described in terms of development objectives. The ideology of development was reinforced by the controlled mass media and the educational system where references to development goals, values and techniques became more common.

Marcos also affirmed a close link between development concepts and the idea of democracy, although he had already expressed this view in his 1969 and 1971 works. He contended that "Democratic institutions must be instruments of national advancement. Democracy must symbolize change,"[4] and that "democracy" has the potential and power to respond to demands for change, indeed that it can act as a catalyst of change. His attitude towards "democracy" was, however, ambivalent. Although he linked it closely to development, he had in mind a particular type of political system. He thought that the connection between democracy and development was not automatic and that only some forms of "democracy" had the revolutionary potential to bring about desired changes. He defined such a system in his concept of a "New Society," which will be discussed shortly.

Political Development

In politics, the new thrust followed an authoritarian prescription for political development. This reflected the position of many modernization theorists who came to terms with the increasing incidence of authoritarian rule in the Third World by arguing that an early commitment to democracy was premature and that coercion might be necessary to speed up the process of transformation from "traditional" to "modern" cultures. Samuel P. Huntington, a well-known modernization theorist, argued that in a stable sequence for political change, institutionalization of new structures and values should precede political mobilization and that considerations of political and social stability should precede claims for greater political democracy. Stability, then, was the criterion for the viability of developing systems. This explains Huntington's assertion that the most important distinction among political systems is not their form of government but the extent to which their governments can rule effectively.[5] In the Philippines under martial law this theory was reflected in Marcos's emphasis on social and political stability as the principal goal of government, his advocacy of strong and efficient government, his readiness to use coercive methods to achieve his objectives, his attempts at developing administrative and economic capacities to meet new demands and goals, and in similar commitments aiming at enhancing the authority of the central government.

The thrust of the new political orientation was towards the centralization of power and the rationalization of systems to better integrate their functions and so strengthen their efficiency and effectiveness. The impression of political "dynamism" comes from the adoption of integrative goals and strategies; however, it will be shown later that the old tendency to political and administrative fragmentation emerged in another form under the new dispensation.

Marcos's preoccupation with government centralization and rationalization was a reaction to the weakness of the Philippine State. This weakness was structural: the wide dispersal of social and political power throughout the country militated against national unity and effective national action. There were many local and regional loci of power and the base of national politics was the local constituency. Politicians' loyalty was essentially parochial, and particular dynasties dominated local social and political life. The practice of bargaining politics encouraged political deals and militated against party discipline. In fact there were no real differences between the two leading political parties; the parties were based largely on opportunistic alliances of regional political warlords. Under this system, attempts at increasing central power were strongly resisted. Philippine congressmen rejected attempts to introduce comprehensive national planning and allowed increased central control only in a few areas of lesser importance, undermining the ability of the state to respond effectively to demands for more rapid national development.

At the national level, a number of structural changes were institutionalized under the rubric of "constitutional authoritarianism." Executive power was placed in the hands of the incumbent President who could issue decrees which had statutory force. His decisions were exempt from judicial review and no definite limit was set on his tenure of office; this power was left to his discretion. There was no mechanism of checks and balances to prevent abuse of power, no separation of powers, as in the old government system. Instead, there was, in the phrase of Perfecto V. Fernandez, "a union or concentration of powers."[6] No change could occur in the Administration except by the will of the president, for regular elections were not specified. This "constitutional dictatorship" (Clinton Rossiter)[7] offered no effective legal remedies against invasion of traditional rights and liberties. The Bill of Rights was now without legal force.[8] As Arturo M. Tolentino, the former senate president, put it somewhat euphemistically, the system was "a highly presidential system, with an extremely strong executive."[9]

The Constitution arrived at by the rump Con-Con in 1973 allowed for the eventual emergence of a parliamentary system out of the presidential dictatorship; this was given constitutional force via the various transitory provisions. Institutional continuity was to be provided by President Marcos, who would become prime minister in the new system. The prime minister would be a strong figure combining the old presidential powers with those of the leader of the majority party in the parliamentary system. It was suggested that under this system political parties would behave in a disciplined way and represent definite political programs. This, Marcos claimed would ultimately lead to the creation of a system that would enjoy strong political powers and would maintain control.

Among the martial law changes was a series of reforms which were meant to destroy the "old" political structures and practices and strengthen the central authority and control. The old oligarchy, Marcos promised, would be eradicated. Congress was rendered inactive, activity of political parties was repressed, and freedom of expression and freedom of the media were denied or subject to government control. Popular elections were replaced by controlled plebiscites whose claim to represent public opinion was meaningless because of the absence of opposition. The representative system became essentially an appendage of the government system and political "representation" operated as a camouflage for a one-party system.

At the barangay level, reforms followed a similar pattern of centralization and rationalization ostensibly aimed at "integrating" these units into national policies, planning and development. Although hailed by Marcos as a step towards "democratization of Filipino politics," these reforms, on the contrary, led to a weakening of the barangay system. The reformed barangay system was not allowed to operate freely. The position of barangay leaders was held at the discretion of the president and the system was treated essentially as an administrative arm of the government used for the delivery of public services at village level.[10] Contemporary observers listed a variety of activities performed by this system; they included implementation of government programs (e.g. for energy conservation, family planning, reforestation, nutrition, health and education), local beautification campaigns and the promotion of peace and order. (The last was often represented by the barangay *tanod* or brigades, the local guardians of security which sometimes repressed local dissent to Marcos's regime.) One observer summarized the role of this system by stating that the barangays were "primarily transmission belts of the national government rather than independent sources of policy inputs and feedback."[11] Thus there was strong central control and manipulation of local government. The barangay system was used not only as an instrument of government control over local development but also for mobilizing people for national development, and to induce ideological conformity. These observations have to be qualified, however. In the Philippines government policies are often confined to government regulations but are ineffective in practice.

A similar trend was discernible in local and regional government units. Despite statements by national leaders in support of enhanced local autonomy, students of local institutions discovered "a recognizable trend to centralize again at the lower level of government" and that the partnership concept "by which the national government guides but does not control . . . [and] both levels of authority can act as effective partners in socio-economic development does not work under the prevailing conditions."[12] A similar point was made by Belinda Aquino, who stated that reforms instituted in local administration did not increase the power of local government units; rather, they tended to streamline governmental structures and maintain central direction of various programs.[13] Raul de Guzman and A. G. Pacho in their 1975 study on central-local government relations, also concluded that "centralism has continued to increase rather than diminish."[14] The same trend was followed in government planning. The various reforms aimed at regional decentralization did not necessarily increase the autonomy of regional political or administrative units; rather, through NEDA's regional offices, they extended the hold of national planners over the economy and brought about increased centralization of power. The administration of regional development remained within the framework of national planning, and the interest of central planners was mainly in maximum utilization and mobilization of regional resources for national development. In the comment of one of the chief regional administrators, "region-oriented development" projects (that is projects generated independently within the region) "appear to be a product of coincidence or accident rather than the product of organized planning"[15] and the relationship between national line agencies at regional levels and at national levels involved "still uncoordinated and unintegrated planning and programming processes." He added that "at best . . . the RDC [Regional Development Council] is still considered by many agencies merely as a forum for discussions."[16] This administrator traced the "basic problem" of regional decentralization to the fact that regional programs and budgets were "prepared and decided upon wholly by the Central Office," which "defeats the purposes of the decentralization of government." He also remarked that there was a "lack of delegation of authority."[17] During this period the local penetration by central authorities increased considerably as the activities of government line agencies expanded both in urban and rural areas.

Apart from these reforms Marcos sought to strengthen the state by reorganizing the political system along corporatist lines. The old pluralist model of representation was to be replaced by a system of representation based on an "organic-hierarchical paradigm," in which all major functional sectors of the society were to be incorporated into the state "under the direction of the central administrative apparatus, and, thus, eliminating 'wasteful competition and internecine disorders bordering anarchy'."[18] This resulted in a hierarchy of functional organizations culminating in one all-encompassing organization at the national level attached to the government system. For instance, each industry had its own labor federation leading into the Trade Union Congress of the Philippines, which would be controlled by the Ministry of Labor. A parallel system was formed for the employers, and both systems were incorporated into the National Tripartite Congress of Labor, Management and Government. Similar structures existed in other occupational sectors. A parallel system was created in politics; it was intended, in Sa-Myung Park's expression, "to replace the old elective representative apparatus with an all-encompassing, appointive representative apparatus so as to channel controlled participation."[19] This corporatist political hierarchy started with Barangay Assemblies in villages, included Municipal Councils and culminated in the National Legislative Council (*Batasang Bayan*) which also included representatives from other sectors. In the view of critical observers, this system of representation was "intended to consolidate bureaucratic control over the society rather than popular control over the state bureaucracy"[20] and it failed as an experiment in system building. Among the reasons given for this were that it was introduced too rapidly, that it was not followed up with sufficient vigor, and that it did not appeal to workers because of a decline in working conditions. It was also increasingly disliked by influential segments of dominant elites who were critical of excessive state intervention and of the active support given to Marcos's crony associates, and who wished to restore the old representative system.[21]

Marcos's ideology of the "New Society" lent additional support to the new thrust at centralizing government power and accelerating the rate of national development. The important element is the notion of positive and active government, which defines Marcos's concept of a desirable political system. Marcos rejected as excessively negative the concept of limited government, which he associated with the Lockean-Jeffersonian "Anglo-Saxon" model of classical liberalism. He claimed that it assumed hostility between the individual and the state, and he denied that there was such a dichotomy. He said:

> Under the Jeffersonian ideal, the entire political system is based on the notion that citizens must ceaselessly defend themselves against the institution of government and that the government that governs least is the best. But now our hope is effective government that can program the growth of the society.[22]

Marcos saw a need for an "effective and active" political system "premised on the active and ceaseless pursuit of national development, human welfare and social justice,"[23] which he contended was necessary for development.

Related to this concept was Marcos's idea of "revolution from the center" and the idea of government acting as an instrument for the "liberation of man." The former envisaged government as an agent of change in transforming the existing social and political order. The latter envisaged an active role of government in bringing about human freedom. In Marcos's phrase, "the government is an institution that liberates, an institution that makes man free."[24]

The other elements of this ideology concentrated on desirable behavior and values. These were embodied in seven " moral postulates"[25]: Filipino Identity, Nationalism and National Unity, Social Justice and Equity, Participatory Democracy, Development and Prosperity, Freedom of Belief, and Internationalism: Oneness with Mankind. Here the ideology was intended to transform, through appropriate reforms, social values towards a more disciplined and responsible outlook and develop a sense of national purpose and unity. The contribution of such ideological appeals seems to have been limited. They are frequently viewed as unimportant for political practice, as reflecting merely the familiar Filipino language of political rhetoric, and as being essentially a cosmetic product of a regime seeking to boost its legitimacy. On the other hand, with their call for " Filipino social and political rebirth" and " a Filipino-style revolution" these appeals nonetheless played an important strategic role in the early years of the new regime and helped to win it some legitimacy. Their presence may also partly explain the widespread popular acquiescence in the martial law regime, or as one writer has put it, " why many Filipinos initially accepted the president's call for reform through authoritarian rule."[26] Whatever their ultimate value might be, these ideas lost their luster in the later part of the Marcos regime. They were not seriously applied in the educational system, were increasingly obviously not lived up to by the First Family, particularly the First Lady, and were increasingly irrelevant to the deteriorating conditions of life of most Filipinos. They were also not adequately integrated with other government ideas on development and government practice. This discussion seeks only to draw attention to the use of ideology to lend support to ideas of accelerated development and to Marcos's favorite political concepts which project a similar outlook. A more direct relationship between this ideology and the government's commitment to technology and science will be examined in a later section.

Administrative Development

The thrust of government administrative reforms was in the same direction as its political reforms. They aimed at centralizing and rationalizing the government system. They were expected to strengthen the administrative capabilities of the national government and lend support to its commitment to more rapid economic growth. A number of administrative reforms were introduced during this period, the most important of which was the review of the entire organizational machinery of government mentioned earlier. The original recommendations of the Commission on Reorganization were published in early 1972 but it was not clear whether they would be accepted in Congress because of its traditional hostility to any extension of central authority, which was now proposed in many respects. However, the situation changed dramatically with the declaration of martial law. The proposed scheme was adopted immediately, becoming the first legislative act of the new administration. It was declared law as the Integrated Reorganization Plan (Presidential Decree No. 1) in September 1972.

A number of other reforms followed the reorganization scheme in a variety of specific areas. Some focused on the creation of new agencies, the abolition of old ones and the streamlining of operations and procedures to improve coordination and performance in government systems. One example was the establishment of the Ministry of Energy in 1979, another, the establishment in the same year of the MAIP (Management Audit and Improvement Program) to "assure a more effective system and procedure" in government agencies. Other reforms were aimed at fostering professionalism in the civil service; as Marcos put it, "professionalization of the civil service should rank high in our priorities." The reforms included programs of in-service training and further education, as well as programs for managers and leaders which utilized recent advances in organization and management theory. Their objective was to improve managerial and leadership standards and competence in handling the problems of modern complex and dynamic public organizations as well as to instill loyalty to the New Society's philosophy of development; in the statement of one government order[27]: " the training program . . . is intended to expose them [future leaders] to concepts and techniques of effective management. These are exemplified in the 'Seven Pillars of Nationhood'." Best known among these were programs for strategically located positions under the Philippine Career Executive system, inaugurated in 1974, the JET (Junior Executive Training) and the STREAM (Supervisory Training for Effective Administrative Management). In addition, the Civil Service Academy of the Philippines was created in 1977 to formalize and implement centralized training programs for middle-management positions in the civil service.

Similar reforms were applied to administrative ethic and discipline. Purges, or threats of purges, of " incompetent or corrupt" officials periodically occurred at virtually all levels of the administrative system.[28] Special courts were instituted in 1978, the Sandiganbayan with special jurisdiction over graft and corruption and the Tanodbayan or ombudsman. Measures were taken to keep administrative expenditure under tight control by reducing cost on personal services, maintenance and operations, equipment, infrastructure and other government services.[29] To reduce red tape, penalties were set for delays in the processing of papers in government offices and, to increase " merit," all appointments by the Civil Service Commission were declared to be subject to the approval of a selection board (on which, in the case of selection of workers for promotion, both the management and rank and file were supposed to be represented " to ensure objectivity").[30] Administrative discipline acquired new urgency in the late 1970s in response to criticism that the gains in discipline made in the early years of the New Society regime were slipping away:

> . . . it is now becoming apparent that much of the headway made in this direction [national discipline] is being lost to a gradual sliding back to the old ways and

> practices not only by the people in government but also by the people themselves. Graft and corruption has again infected the bureaucracy, particularly in the lower echelons where the public is again victim to petty extortion by functionaries who refuse to do their job without grease money. Red tape perpetuates the obnoxious practice. The "tong" system bribing of policemen and other abuses by law enforcers have resurfaced.[31]

Despite the legislation and propaganda used to instill discipline in administrative conduct, comments like these raised doubts about the effectiveness of the reform measures. The new regime, like the old one before, was facing the problem of how to overcome traditional habitual conduct, characterized by such "prismatic" values (as Fred Riggs calls them) as behavioral ritualism and formalism.[32] Nonetheless there was a real government commitment to reforms in public administration. These encompassed organizational structures, behavior and values and aimed at increasing the efficiency and effectiveness of government systems and their professionalism, as well as at inculcating more accountability among civil servants.

Economic Policy

Increased interest was also shown in more robust ideas of economic growth. This was reflected in frequent references to "accelerated development" and in economic reforms and policies introduced during this period. The following discussion is not directly concerned with the progress of the economy as such, but rather with economic concepts and methods used in policies and planning which tended to reflect technocratic values. It also draws attention to the increased role of technocratic planners in policy-making.

The new thrust in economic policy became noticeable with the establishment of the BOI (Board of Investments) in 1968, which was given the responsibility to provide a comprehensive system of investment incentives for private enterprise. The BOI introduced two major innovations aimed at injecting more dynamism into economic policy. The first was an explicit commitment, by instituting an incentives policy, to industrial growth. The incentives offered included tax exemptions, accelerated depreciation and other concessions and were given mostly to large industrial firms, capital-intensive in their operation.[33] This strategy seems to have been based on the belief, supported by many development economists then, that it would provide a strong stimulus to further growth by its linkages with other sectors and so generate dynamic growth in the entire economy. In practice this policy supported capital-intensive projects, and the economic cost and overall economic soundness of investment appear to have been the sole criteria for judging the incentive worthiness of firms that applied for incentives. This policy was summarized in one BOI document in 1970: "The principal incentives it [the act] provides are designed to support capital-intensive undertakings. This," the document added, "appears to be consistent with the general objective of promoting the establishment in this country of intermediate and capital goods industries. The technology in this type of industry is usually capital-intensive and has limited opportunities for labor substitution."[34]

The second innovation was a move towards greater economic rationality which included more extensive planning, greater rationalization of the economy and increased government intervention. Indeed, the possibility was raised that in this respect the Board might be going too far. This was hinted at in 1971 by the spokesman for the powerful American Chamber of Commerce.[35] According to him, the BOI policy sought to merge "competing ventures" to prevent waste of scarce resources and "to wield the big stick against undesired investments, at the same time that it is waving the carrot to desired ones," and its role in national development was "a more active, more coercive, role, something that will enable it to force the issue of rationalization where necessary." But despite reservations by some elements of the private sector, the new interest in rational planning and economic rationalization gained support of a number of powerful government agencies such as the NEC (National Economic Council), the Presidential Staff, the Office of Economic Coordination and the Fiscal-Financial Policy Committee, which were all staffed with the new breed of " professional economists and corporate managers," the new techno-bureaucrats.[36]

The new economic thrust manifested itself more fully in the policies adopted after the establishment of the martial law regime. These followed a two-pronged approach. The first involved the advocacy of indicative planning and later the use of integrated approaches. The second involved a more balanced concept of national economic growth, although the original BOI approach defined five years earlier does not seem to have been abandoned either. The source of this policy orientation was the Ranis Report of 1974 entitled Sharing in Development.[37] Its principal recommendations were: more balanced development in both urban and rural areas, replacement of the old import-substitution strategy by a more comprehensive economic approach, emphasis on export orientation, and the recognition of the need to change the growth path of the economy in a more " employment sensitive" direction, i.e., towards increased generation of employment. A " policy package" was prepared to indicate what concrete reforms should be used to achieve these objectives.[38] Marcos himself, following Ranis, opted for " balanced" or " integrated" approaches in his new plan for Philippine development. A variety of strategies[39] were proposed such as the promotion of export-oriented industries, the encouragement of labor-intensive industries and techniques of production, promotion of backward integration in the industrial sector, and regional dispersal of industries. According to Marcos, " The combination of those policies constitutes in our view a rational and integrated approach to the problem of engendering industrial expansion and general economic development."[40]

The new economic dynamism was to be generated mainly by private enterprise. According to Gerardo Sicat, the head of NEDA, "The programs of the New Society in industry and agriculture are the same ones that were pursued in the old society. But the one important difference is the speed and flexibility with which decisions of government in relation to these programs are now made."[41] As before this period, the economic framework underlying such programs was based on the belief that free enterprise should be the principal engine of economic growth. This was agreed on both by the industrial elite and the technocrats who were in charge of national planning. It was also Marcos's position. In his *Notes on the New Society*, he stated that the Philippine economic system was "basically private enterprise,"[42] and in another passage he called the private sector or free enterprise "a fundamental element of our national life"[43] and "the sole engine of economic growth" and the entrepreneur "the central factor in development . . . a hero of development."[44] Marcos reaffirmed this position in 1978 in the Interim Batasang Pambansa (Interim National Assembly) where he spoke against a proposed nationalization bill, arguing that it "would fundamentally alter the policy we have long maintained, that the government looks to the resources, vision, and know-how of free enterprise for the achievement of productivity and growth in the national economy."[45] He also perceived the bill as contradicting the spirit of the investments policy which encouraged participation of foreign investors in economic development.

The emphasis on private enterprise must, however, be qualified. The economy contained a strong element of public guidance through government regulations and other controls as well as through public enterprise. Government intervention also increased with the expansion of planning and the rationalization of systems. Gerardo Sicat described the prevailing economic philosophy as a mixture of liberalism with planning; he said: "this principle of liberal economic policy with appropriate governmental intervention constitutes the major theme of economic rationalization of the New Society. This theme has been inherited from the economic reforms after the crisis of 1970."[46] Technocrats like Sicat appreciated the dynamism of private enterprise but also saw the need for government intervention both to tame the "irrationality" of this system and to channel its energies in a socially desirable direction.

The new more active role of the state manifested itself in several complementary ways. To begin with, it involved the adoption of policies aimed at accelerating economic growth. The BOI expanded its policy activities from investment incentives to a loan assistance scheme administered by the DBP (Development Bank of the Philippines), which led to a radical increase in the number of assisted projects and in the distribution and dispersal of loans. Vigorous support was also given to foreign investment which often superseded domestic investment.[47] Elsewhere economic growth was to be stimulated through infrastructural programs and projects. Figures cited in the National Development Plan 1978–1982 showed that during the period 1973–1976 the largest slice of government expenditures went to economic development (37–45 percent of the total budget) and that this trend was to continue during the following decade (35.8 percent of the total budget).[48] In this area infrastructural activities were given priority; they included transportation, water resources, power and electrification, communication and "social infrastructure." Sicat observed in 1980 that this priority "will remain the mainstay of budget together with a promise to accelerate productive momentum in agriculture, energy development and industrial growth."[49] The new emphasis on energy generation was necessitated by the dramatic rise in the price of crude oil. Also, plans were put under way to set up export-processing zones which would, it was hoped, attract foreign investment, thus increasing the country's exporting capacity and generating employment.

A more active role of the state was also reflected in the area of fiscal and budgetary policies.[50] Now the utility of programs rather than conventional budgeting methods was to be used for judging the desirability of fiscal and budgetary policies; i.e., programs should be measured by the yardstick of their expected contribution to economic growth. In the statement of one National Plan, "there has been a constant effort to make the budget a more viable tool of economic development as reflected in the budgetary themes of dynamic flexibility and self-reliance adopted in the past two years [before 1977]. Budget preparation and implementation processes have improved to be more responsive and supportive of development objectives."[51] Another significant innovation was the decision to make the approval of the national budget the first item on the agenda of the national legislature, while in the past it had been more or less the last. Similar innovative reforms occurred in fiscal policy.[52] They included a revision of the basic tax codes (designed to generate more revenue and improve efficiency in tax administration), increased taxing powers for local governments and increased real property tax. The national budget was placed under the control of the DBCC (Development Budget Coordinating Committee), which included representatives of all central economic agencies or ministries; this aimed at harmonizing annual budgets with development planning. In this way the process of planning and budgeting were institutionally linked. Another innovation occurred with the implementation of such advanced techniques of accounting as zero-base budgeting.

The new economic thrust towards centralization and state intervention also manifested itself in another area which came to play a central role in national development, namely national planning. One central agency was created, NEDA (National Economic and Development Authority), which, according to Sicat, who became its head, was to be "the principal coordinative body for economic and social programs of action" and aimed at "a faster and more efficient process of decision-making."[53] It was staffed with some of the country's leading technocrats. The establishment of NEDA made possible for the first time an "integrated approach" to development activities. As Marcos put it, "we brought together under one roof the whole apparatus of economic planning and management in the country."[54] The importance of NEDA was highlighted by the fact that the president himself was the chairman of its board. The "philosophy of NEDA" made itself increasingly manifest in the Philippine economy. It included a commitment to increased government economic intervention and nation-wide "indicative" planning (so-called because it "indicated" the proposed direction of development) as well as interest in the rationalization and integration of systems. These commitments led to a strong dose of manipulation of the economy with NEDA guiding or pushing the course of economic development in the desired direction.

This period witnessed an increased role for technocrats in policy-making and planning. They gave impetus to the ongoing economic and administrative reforms and took a leading role in the debates on the future direction of the economy. In the government they were represented by a trio of ministers: Gerardo Sicat, the head of NEDA and minister of Economic Planning and Policy, Roberto Ongpin (Industry) and Cesar Virata (Finance), who later became prime minister. These leading technocrats took the "transnationalist" position, in favor of liberalization of imports, removal of tariffs and export-oriented industrialization, in contrast to the "nationalist" position of the advocates of Filipinization and local protectionism. In this respect, they allied themselves with economic policies advocated for the Philippines by the World Bank (WB) and the International Monetary Fund (IMF), which wielded enormous power over the Philippine economy by their ability to dispense (or withdraw) external credit urgently needed for economic development or, especially later, for bailing out the government in times of financial crises.[55] In this way, Philippine technocrats, helped (some say guided) by these multilateral agencies, had a decisive impact on the economy, particularly during the later 1970s and early 1980s.

In summary, government commitments to innovative economic reforms and policies during the early period of the Marcos regime manifested themselves through new economic strategies, the new economic activism of the state, and the introduction of comprehensive planning, which were expected to "accelerate growth" in the national economy. The organizational thrust was towards centralization of authority and the rationalization of systems. The principal role in this process was to be played by the private sector, but the state's intervention was to be strong and to be exercised through government planning. Technological and technocratic ideas were deployed in this process and the influence of technocratic administrators in policy-making assumed crucial importance. In spite of this trend, the economy started sliding in the late 1970s and continued to do so during the remaining period of the Marcos regime. There was a dramatic deterioration in terms of trade which in turn, led to a debt crisis; there were increasingly monopolistic practices in critical agro-industrial areas which tended to rigidify the domestic economic system; there was also a lack of coordination in policy-making despite the planners' emphasis on integration,[56] and the tendency to pillage the system through " crony monopolies" and otherwise. This tended to undermine rational control over the national economy and contributed to its gradual decline.

Social Policy

There were various government initiatives during this period in the area of social policy. They were reflected in proposals for policy reforms and planning and in Marcos's political rhetoric. The preoccupation with social policies was not new. Commitments to social objectives were specified in the Philippine Constitution of 1935, which declared " the promotion of social justice to ensure the well-being and economic security of all the people" to be one of its principal goals. However, during the liberal era, interest in social policy was limited and social development was considered as more or less a by-product of economic growth, with which it was perceived to form " an indivisible duo."[57] Little was done in the 1960s or before to promote social policies, except for the traditional support for education. Also the Magna Carta of Justice and Economic Development of 1969 was not a document that would by itself bring about radical changes in Philippine social development. Rather, it was an ideal statement of intentions which drew on the tradition of appeals to social justice for the Filipino "little man" going back to President Quezon in the 1930s. But it was not much of a blueprint for political action.

An early attempt to relate the "social welfare" factor to the new thrust towards accelerated development followed the establishment of the Board of Investments. The cornerstone of the BOI approach was the factor of "social desirability" as a criterion for evaluating the worthiness of projects considered for government incentives. Yet, as Ocampo pointed out in 1971,[58] this criterion made little difference in practice and did not achieve the desired objective, which was a more equitable distribution of income. For labor-intensive investment, generating employment, which should have been considered under this criterion, tended to be neglected when it had to compete with the industrial linkage criterion. Ocampo described this situation as "frankly encouraging capital-intensiveness in investment."[59] This policy concern was, however, strengthened later as the changing priorities of BOI's investments plans indicate. The Sixth Investments Plan explicitly stated that preference should be given to small- and medium-scale industries, although certain qualifications were added.[60] Also, a new criterion for measuring labor intensiveness was adopted which aimed at maximizing employment opportunities created by new investment.[61] The BOI Export Priorities Plans which followed after 1975 went further by making the creation of labor, not mere expansion of production, one of their priority objectives. Labor intensiveness and the contribution of projects to balanced regional development were also used as criteria by some government loaning agencies (e.g. the IGLF, or Industrial Guarantee and Loan Fund).[62] Thus the "social welfare dimension" of economic activity was to be given recognition. However, it is obvious that these BOI strategies contributed little to social equity in government policies and to more widespread improvement in the people's lives.

Nonetheless, during the early martial law period, attempts were made to relate issues of social policy to national planning. Now the old social strategies and concerns with agrarian reform, cooperatives, employment generation and community development were projected by Marcos as "devices for galvanizing national economic capabilities and for spurring the national economy into new feats of productivity."[63] They were more integrated with national planning and subject to tighter central control. Figures for the period indicated a considerable shift in resources towards the "social sectors," although it is not easy to determine whether this led to a major improvement because the term had been periodically redefined by planners to project government social concerns in a more favorable light. They showed that during the period 1973 to 1976, between 19 and 22 percent of the national budget was to be spent on these sectors and that by 1987 this allocation would rise to 31 percent.[64] Figures also suggested substantial increases of benefits in a relatively short time in areas such as health and nutrition, education and manpower, housing, social services and community development[65]; life expectancy was expected to rise from 60 years in 1976 to 64.4 years in 1987.[66]

Marcos's preoccupation with social policy had a strong ideological dimension. His message was that people should not be mere recipients of benefits but should actively participate in their production. He also emphasized the need for self-reliance and self-development. He claimed that the purpose of social policy was not merely charity; rather, it was to help people help themselves, to make them productive and encourage self-reliance. As much as possible they should depend on their own resources and skills for improving their existence.

Marcos's rhetoric of social development also embodied "humanist" references to the well-being of man and a better quality of life. He defined the humanist approach as "the improvement of the quality of life of every Filipino"[67] and identified it with several objectives. The first was equality of opportunity, which was taken to mean the equalization of opportunities rather than absolute equality. The second was "democratization of income and private property." This was intended to highlight the social function of property rights. These, Marcos declared, "are regulated to the extent that their use and the enjoyment of their fruits serve social no less than individual purposes."[68] The third objective was the attainment of social justice. Marcos identified this concept with equality of treatment and a fair share of benefits, and contended that it was " part of the government's attempt to create a just society, a society where opportunities for advancement are not a monopoly of the privileged few but are made available to all sectors of the national community."[69] He also linked social justice to the need to overcome social disparities and strengthen the middle classes, insisting that the " wide gap between the rich and the poor be narrowed."[70] Humanist and economic development are not separate however; they go together. This message was reflected in Marcos's proposed strategy for the 1980s, which he described as two-pronged: " One is the attainment of a dynamic and balanced economy. The other is the provision for a more equitable access to social development."[71]

Numerous schemes were introduced to translate humanist ideas of development into a set of public policies. They were intended to improve the condition of the poor, and included construction of houses for lower-income people and the eradication of squatting areas and resettlement of squatters. Another scheme was the Samahang Nayon cooperative movement which advocated self-reliance, people's active involvement in projects, greater equalization of income and the broadening of the middle-class base. The best known institutional arrangement was the Ministry of Ecology and Human Settlements (MEHS) established in 1978. Hailed as representing " the humanist ideology of the new order,"[72] this ministry sought to extend the commitments of the New Society ideology to the area of social development. It was given the task of providing " adequate and decent standard of living for all Filipinos" in housing, health and sanitation, education, culture and other social aspects, and for that purpose to coordinate and monitor the relevant activities. It had also three other duties: to protect the environment and maintain ecological balance, to develop the shelter system, and to utilize land properly through planning and zoning. This powerful ministry was placed under the First Lady, who appears to have perceived herself as a missionary in this role. It was housed in prestigious buildings, received strong political and financial backing at the national level, was supported by the popular press, and enjoyed, initially, considerable appeal. Staffed with young technocrats, it was a very busy place both at the headquarters' level and in its regional or local offices.

At the center of the ministry's approach to social policy was the articulation of eleven "basic needs." These included water; food (agriculture); clothing (cottage industries); economic base (livelihood); medical services; education; culture and technology; ecological balance, sports and recreation; shelter (housing and land use); and mobility (transport and communication).[73] It was contended that this approach was "total and integrative"[74] and aimed at satisfying both material and spiritual needs. A package of services was to be provided to help people meet their basic needs; however, it was proposed that the government should provide only "the minimum whole," the basic services; this was expected to motivate people to further effort, to achieve "the maximum whole" for themselves.

Several strategies were based on this approach. The better known among these were: the renewal of communities in depressed or squatting areas (e.g. the Tondo Foreshore Urban Development Project),[75] construction of model communities or barangays in new areas, creation of strategically located growth centers, resettlement projects and the BLISS (Bagong Lipunan Improvement of Sites and Services) program. In this program the BLISS planners claimed to apply a "holistic" approach to community development, specifically to projects which aimed at upgrading the environment in slums or meeting the housing needs of lower-income people. In one such scheme slum residents were to be provided with basic services—water, lighting, sanitation, good roads. This was expected to motivate them to further improve their surroundings on the principle of self-help.[76] In another scheme, residents of BLISS apartments were to be taught livelihood skills to enable them to develop self-reliance and create independent, self-sufficient communities; as Mrs. Marcos put it, "to attain a measure of economic independence and an ability to break out of the poverty cycle into more productive, dynamic citizenship."[77] Another well-publicized scheme aimed at helping to generate new opportunities for economically weaker businesses. This was the "Big Brother" project of the First Lady, in which big private corporations were invited to make available their capital base, technology and expertise, as well as their business connections, to rural-based and medium-scale industries. This project was also sponsored by powerful business interests (e.g., the Filipino Chinese Chamber of Commerce) for job-creating training programs.[78]

Despite powerful political and financial backing, the Human Settlements scheme did not make much impact on social conditions for the poor. Like the other aspects of the New Society ideology, it ran out of steam and seems, eventually, to have had a negative influence. The schemes broke down in grandiose projects that wasted money, ostentatious public relations, arbitrariness in decision-making, corruption among its officers and the erratic leadership provided by the First Lady. Its projects were too dispersed and too few for a large and populous country like the Philippines. Most Human Settlements projects were undertaken in Metro Manila, where they enjoyed glossy publicity, confirming the impression that the type of development aimed at was cosmetic rather than real. Programs aimed at squatter communities like the Tondo foreshore were more often than not hollow facades erected on the ruins of real self-help initiatives destroyed by martial law.

Considered as a whole, the area of social policy manifested initiatives which were intended to support the regime's thrust towards accelerated development, but were largely rhetorical, particularly during the later period of the Marcos regime, when there was a sharp decline in living standards for most Filipinos. The strategies and approaches applied in these policies followed models devised by technocratic planners and were treated as an integral part of development planning. Even the "Human Settlements approaches," ostensibly reflecting "humanist" concepts, were developed by the technocrats of the MEHS who were determined to do their own social sectoral planning.

* * * * *

This chapter has sought to examine the claim to an increased "dynamism of development" in the Philippines during martial law. In politics, this trend was mostly represented by the authoritarian regime's use of its dictatorial powers to deal with immediate problems of social unrest and disorders and to strengthen its power. In government administration, commitments to reform were initially strong and pervasive and aimed at maximizing the efficiency and effectiveness of systems. A similar trend manifested itself in economic policies and included the establishment of NEDA as the sole central planning agency. The state also became very active in the economy, becoming perhaps the dominant force. In the area of social policy "social sectors" became an integral part of national development and planning. It was given prominence in the rhetoric of President Marcos, who defined it as "improvement in the quality of life" and sometimes referred to it as "humanist development." It was related to the New Society ideology, which it provided with a social content.

This policy orientation reflected the new thrust in development planning towards "accelerated growth." In this orientation the approach to policy reforms involved the use of models and ideas that were essentially technocratic in nature and included comprehensive planning, rationalization of systems and centralization of decision-making within the broader context of national development planning. Thus the "new dynamism" in policy-making and technocratic approaches were closely related and the technocratic approach became the dominant mode for dealing with public policies. This ascendant technocratic influence will be illustrated in the case studies of the next chapter.

CHAPTER IV

Legislation, Projects and Plans

The preceding discussion underlined the dynamic context of Philippine development planning during the martial law period. This was reflected in attempts to accelerate the rate of national growth and in the use of models and approaches with a technocratic origin. Chapter IV illustrates this trend through a number of short studies based on events which occurred in the Philippines in the 1970s in different areas of government activity: legislation, development projects and development plans. Specifically, it seeks to identify the extent of technocratic influence in these areas by examining the cases for the presence of (1) a change in development perspectives in a technocratic direction; (2) the choice and use of concepts, models and techniques reflecting a technocratic outlook; (3) the orientation of administrative systems and administrative conduct in this direction.

Legislation for Development

This section surveys Philippine legislation in the 1970s in the area of natural resources, specifically forest and watershed conservation and development. Six legislative acts pertaining to forest and watershed regulations are examined for the presence of technocratic values. They are examined in terms of (i) their policy objectives, (ii) the approaches, concepts and techniques applied in their implementation, and (iii) the treatment of people exposed to this policy and how the acts proposed to deal with human problems arising out of their implementation.

At the start of the 1970s, forest and watershed areas in the Philippines were regulated under Forestry Administrative Order No. 11 (Revised) of the Bureau of Forest Development (dated September 1, 1970).[1] This code, the "revised forestry license regulations," embodied an essentially administrative approach, covering policy in a broad way. It dealt with matters of both technical and human interest, including "policies on public forests, alienable and disposable land, private rights, limitations of concessions, limitations of areas and diffusion of ownership." It seems mainly to have been intended to regularize and improve the administrative machinery of the existing system.

In this code, forest and watershed areas were declared to be a matter of public responsibility, and commercial utilization of public forests was disallowed "unless such grant or authorization advances the economic and social welfare of the Filipino people while at the same time assuring the continuity of the forest in productive condition."(Ch. I, 3a) It was forbidden to cut and remove timber except by public bidding, although this requirement could be modified on grounds of "public interest," when "vacant forest area" could be awarded by the Director of Forestry. (Ch. 1, 3a(1)) Alienation or disposal of forest lands was subject to restriction, and the private rights of traditional residents were explicitly recognized. There were references to "rights of ownership under existing laws and, in the case of cultural minorities, to rights of possession existing at the time a license is granted under this Order, which possession may include places of abode and worship, burial grounds, and old clearings, but excludes production forests . . ." (Preliminary Chapter 2 ee) Explicit recognition of private rights was also given in the following provision of the code:

> The grant of any license under this Order shall be subject to the private rights of cultural minorities and of any other person, if there be any, within the concession or licensed area, as evidenced by their occupation existing at the time the license is issued by the government, or other nuniments of title, and the area over which such private rights exist shall be deemed excluded from the concession or licensed area, and

logging operations within said reservation or area shall be allowed only if covered by express authorization from the licensing authority. Whenever resources of the Bureau of Forestry permit, such private rights shall be identified by the Bureau of Forestry, or when resources of the Bureau are inadequate, by the prospective licensee as authorized by the Director of Forestry, prior to the grant of any license. (Chapter 1, 3c)

In these provisions of the code, on the one hand, traditional rights were given recognition; on the other hand, the " public interest" and the advancement of " the economic and social welfare of the Filipino people" could be invoked as a justification for granting concessions and issuing licenses for commercial logging. The Bureau of Forestry was to act as the licensing authority. This important point was not, however, given particular emphasis, although it justified large-scale state intervention under the guise of public interest.

Apart from such reservations, the provisions of the code indicate essentially regulatory-administrative interests. An illustration of this was the provision seeking to accommodate the needs of customary occupants of the land. The code stated that " gratuitous licenses" are provided for, free of charge, " for personal use," entitling holders " to harvest and utilize forest products . . . in reasonable quantities."(Ch. XI, 51c, and Ch. VII, 27d) Furthermore, watershed areas critical to the water needs of small communities were allowed to be " locked-off from logging and all forms of exploitations by specific action of the Director of Forestry when in his opinion alternative conservation measures are not sufficient."(Ch. X, 49b (6)) Another illustration is the provision allowing poor residents to use forest products " for their personal use where no communal forest has been established for them."(Ch. XI, 51a, 2(b)) A similar concern was shown for the conservation of food resources upon which the residents of forested zones partly depend. The code declared that

> It shall be unlawful for any person, not in possession of any special license issued by the Bureau of Forestry, to cut from the public forest, forest reserves, national parks or communal forest, for use as lumber or firewood any tree or bush, the fruits of which can be used as food for human beings. (Ch. XI, 53r)

The last provision was, however, in conflict with the provisions of the code granting the enjoyment of land to the cultural minorities. Now the " owners" of the land are not allowed to " log" it without the government's permission.

The code also gave explicit recognition to communal ownership of forest land. " Communal forests" were defined as tracts of public forest set aside for the " exclusive use of the residents of a municipality from which said residents may cut, collect and remove forest products for their personal use in accordance with existing laws and regulations." (Preliminary chapter, q) Such communal forests were, of course, vital to the livelihood of cultural communities, many of whom depended partly on hunting and gathering for their sustenance and on various minor forest products for local handicrafts, household needs and production.

The second legislative act pertaining to forest and watershed areas was the Presidential Decree issued four years later (P.D. 384). Although it did not depart dramatically from the code of 1970, this decree introduced a more self-conscious managerial outlook. It claimed that there was a need for more energetic national policy in the forestry and watershed sector: " there is a serious need to effectively conserve the Nation's public forests, including watershed areas, to protect and preserve national parks, and at the same time to provide suitable agricultural land for our people." No major change in policy was apparently contemplated in regard to the people affected by the code, although their number, based on an " official population census," was persistently downplayed.[2] This is evident in Section 37 where it was stated that

No licenses, leases and permits, however, shall be granted in provinces and cities which, according to the latest official population census, are inhabited by members of the national cultural minorities.

The new management element was reflected in the proposal to "integrate" the administration of forest areas. It also included proposals for relocation. Occupants of forested zones, who were now referred to as "primitive tribes," were to be resettled, ostensibly to integrate them into the "socio-economic mainstream" of the Philippine society. This element was present in two sections dealing with *kaingin* management:

> The Bureau . . . shall identify and stabilize the land claims of primitive tribes [this refers to swidderers rather than terrace farmers] residing within the public forest. Provision shall be made for permanent settlement on designated areas reserved for the purpose, such as unoccupied alienable or disposable lands, relocation centers within the public forest and elsewhere as may be required to achieve the objectives; Provided, that a member of the primitive tribes who, since July 4, 1955, has continuously and publiclly occupied and cultivated, either by himself or through his predecessors in-interest, a tract or tracts of land declared as alienable or disposable. (Section 34)

> Resettlement or integration of settlers in public forests through a system of kaingin management. This involves a complete and continuing census of all occupants of public forest, including their location, in order to determine valid claims and prepare action plans for integration of such occupants into the socio-economic mainstream . . . (Objectives in the Foreword (6))[3]

The third legislative act passed in this area in the 1970s was Presidential Decree 705 of May 19, 1975, also known as the "Revised Forestry Code of the Philippines." This increased the emphasis on systems efficiency and focused on effective management rather than the mere "improvement" of administration. "Development" rather than "conservation" of resources became the focus of legislative interest. According to the authors of this code, current laws and regulations governing forest land were "not responsive enough to support re-oriented government programs, projects and efforts." They proposed to bring them into harmony with the government's thrust towards accelerated development through more integrated control over natural resources and the use of advanced scientific techniques. The general objectives of this policy, stated in the introduction, were: the "proper classification, management and utilization of the lands of the public domain to maximize their productivity to meet the demands of our increasing population." The policy centered on land use and management; it included the multiple use of forest lands (which "shall be oriented to the development and progress requirements of the country, the advancement of science and technology, and the public welfare"), classification and survey of land (which "shall be systematized and hastened"), encouragement and rationalization of wood-processing plants, and the protection, development and rehabilitation of forest lands ("to ensure their continuity in productive condition").

The management-oriented outlook of the code was strongly noticeable in references to the treatment of people residing in the affected areas. Although the original definition of "private rights" had been retained (Section 1(mm)), these were not explicitly invoked in the subsequent text. It is also significant that no reference was now made in the definitions of "lease," "license" or "license agreements" to the rights of cultural communities; leases and licenses were now treated as "a privilege granted by the State." As in the preceding document the mentality of the "modernizers" who drew up this document was revealed in the definition of a "primitive tribe"(Section 3(11)) as "a group of endemic tribes living primitively as a distant portion of a people from a common ancestor." Cultural minorities were now lumped in with *kaingineros*, squatters and other quasi-illegal occupants and residents of forest lands, and a complete census was promised " showing the extent of their respective occupation and resulting damage, or impairment of forest resources." (Section 52)

Many other rights formerly enjoyed remained either unmentioned or were restricted. For instance, the Revised Code did not mention " communal forests" or " communal pasture lands," both of which were vital components in the subsistence economy of many minorities. It also restricted the use of forest products to permit holders:

> Any person who shall cut, gather, collect, or remove timber or other forest products from any forest land . . . without any authority under a license agreement, lease, license or permit, shall be guilty of qualified theft . . . (Section 68)

Gone too were the original provisions for " gratuituous licenses" and the freedom of " poor residents" to extract forest products for personal use. More importantly, there was no longer explicit provision for watersheds critical to the needs of small communities to be preserved from the harmful effects of logging and other forms of forest exploitation. These included such " critical watersheds" as drainage areas, which might support large structures like hydro-electric power projects and major irrigation works.

Despite the many new restrictions on forest occupancy, historical occupancy of cultural communities was still given some cognizance and protection. One provision (Section 15) gave recognition to " open, continuous, adverse and public" occupation of land for a period of not less than thirty years prior to the effectivity of the code. A similar recognition was given to other " well-established communities." The code also stated that

> Kaingineros, squatters, cultural minorities and other occupants who entered into forest lands before the effectivity of this Code, without permits or authority, shall be prosecuted: Provided, that they do not increase their clearings; Provided, further, that they undertake within two (2) months from notice thereof, the activities which will be imposed upon them by the Bureau in accordance with a management

plan calculated to conserve and protect forest resources. (Section 53)

Elsewhere, provisions allowed that

> Wild life may be destroyed, killed, consumed, eaten or otherwise disposed of, without the necessity of permit, for protection of life, health, safety and property, and the convenience of the people. (Section 55)

The Revised Forestry Code of 1975 reflects the changing government orientation in the administration of forests. The emphasis in the policy was on efficient forest management in the context of national planning and development. It was stated that "Forest occupancy shall henceforth be managed" and that any occupancy resulting in sedimentation, erosion and reduction of water yield to the detriment of community and public interest "shall not be allowed." (Section 51) In this policy context, local people were treated rather marginally. In fact the code was intended to make people fit into the new development-oriented outlook. The original commitments to private rights were diluted and the right of abode in forest areas was no longer absolute, as before.[4]

Two subsequent legislative acts further strengthened the efficiency and development-oriented approach to natural resource management. The first of these was Presidential Decree 1559, dated June 11, 1978, amending the Revised Forestry Code, "to further strengthen the Code to make it responsive to present realities and to the new thrust of development policies and programs on forest development and conservation and rationalization of the wood industry." Two innovations were introduced in this decree. One was a commitment to agro-forestry—"a sustainable management of land which increases over-all production, combines agricultural crops, tree crops and forest plants and/or animals simultaneously or sequentially, and applies management practices which are compatible with the cultural patterns of the local population."(Section I,v of P.D.1559) The other innovation was a call for private-sector participation "in forest management, protection and development as well as in wood processing activities within the concept of joint or co-management of forest resources." These innovations were justified as a necessary reorganization of the industry "using management principles, systems and procedures to attain stability, efficiency and profitability of operation." (Section 1, kk)

The rights of cultural communities shrunk greatly in the decree because now it was explicitly specified that "productive forest areas" were excluded from free use. In Section 1(m) the right of possession by the cultural minorities "may include places of abode and worship, burial grounds, and old clearings, but exclude productive forest inclusive of logged-over areas, commercial forest and established plantations of forest trees of economic value." Indeed, regulations controlling local people were now more strict than before. For instance, no utilization and occupation of forest and grazing land would be permitted without a license (Section 20), nor would permission be given to cut, gather and collect timber without license (Section 68). The threat of eviction was invoked for "unlawful" occupation covering all those who enter or occupy any forest or grazing land (Section 68), and subject to penalty for "illegal occupation" were those who without license "hurt, capture or kill any kind of bird, fish or wild animal life within the area in the national parks system." (Section 71)

The strengthening of the management element and the weakening of private and traditional rights was also reflected in Section 20 of the decree which effectively left decisions on the use of forests to the discretion of the administrator acting in "the national interest":

> . . . when the national interest so requires, the President may amend, modify, replace, or rescind any contract, concession, permit, license, or any other form of privilege granted herein . . . the President may . . . order the summary suspension of any such contract, concession, license, permit, lease or privilege granted under this decree for violation of any of the conditions therein such as those pertaining but not

limited to reforestation, pollution, environmental protection, export limitation or such conditions as are prescribed by the Ministry of National Resources in daily issued regulations.

Dated on the same day, Presidential Decree 1581 tightened up P.D. 1559 in relation to grants of government contracts, to "guarantee close administrative supervision and control over contracts" (entered into by any government agency). It prescribed "certain standard conditions for government contracts, concessions, licenses . . . and similar privileges involving the exploration and development of natural resources." (Introduction) The provision of P.D. 1559 invoking "the national interest" was now extended to all government contracts and the President acquired the power to "review, amend, modify, revise, rescind or revoke any such contract" and "summarily suspend and order the cessation of all activities and operations under the contract . . . for violation of any of the conditions or provisions therein or any condition prescribed in duly issued rules and regulations of the administering government agency or instrumentality." (Sections 1, a, b)

The last legislative act passed in the 1970s was a watershed reservation regulation from the Ministry of Energy made public in January 1979 under the title "Proposed Integrated Management of Resources in Watershed Reservations."[5] Even more than the preceding acts, this regulation sought to assure maximum rationalization of resources and aimed at maximizing the speed and efficiency of their utilization. In this act the technocratic-manipulative outlook, which had become increasingly evident from 1975 onward, now became dominant.

As the title of the document suggests, the act reflected comprehensive ideas of resource utilization and planning and made persistent use of such favorite technocratic concepts as "integration" and "coordination" in the context of management and organization systems:

> The watershed reservations shall be managed (conserved, protected, developed and utilized) in line with

> the concept of coordinated or multiple use approach, that is, the resources shall be managed under one central authority for their productive, protective and amenity values. (Section 2 (b))
>
> Watershed projects such as dams, mining, pasture, range, roads and agriculture, among others, shall be evaluated by means of the systems approach with the affected areas as focal points of analysis. Possible adverse effects of the watershed projects on the environment shall also be considered. (Section 2 (d))

The management approach was clearly reflected in the definitions in Section 2, which suggested a technical and professional way of thinking, a comprehensive approach, and a structural and manipulative view of reality. An example was the use of the concept of "integrated management of watersheds" which was defined as "a unified or integrated manipulation or control of all watershed resources to carry out planned activities aimed at attaining specified objectives." Another was the reference to "systems approach" which was defined as "the analysis of problems and formulation strategies which consider all significant components of the system in their interrelationships to attain specified objectives." Other terms for strategies used reflected a similar outlook: "optimal mix of products and services," "agro-forestry," "land-use allocation," "kaingin management," "resettlement," "water production," "geological exploration," "aquatic resources" and "environmental management."

In this act, local people came into the picture more or less only indirectly, as "valuable assets." They were treated as one of the resources to be utilized for efficient management purposes. This is evident from the first statement of the basic policies:

> Watershed reservations shall be managed for the public interest with maximum active participation of the people in the watersheds and adjacent areas. The people in the watersheds are valuable assets which can contribute meaningfully in the efficient management of watersheds, thus, all possible alternatives along

this concept must be examined for feasibility and viability. (Section 2a)

The lives of forest dwellers were to be regulated in the interest of production, putting in question the former commitments to their right of residency. Many illustrations of this are present in the act. The definition of "license, lease and permit" made no reference to the traditional prerogatives of forest residents; it consisted of a simple legalistic statement that this meant "formal or legal documents authorizing, directing and controlling specified activities." (Section 3(g)) The definition of "resettlement" carried a similarly stipulative ring; this was referred to as "the process of relocating human and other resources to more suitable areas for the purpose of improving the quality of life and promoting the conservation of land and other resources." (Section 3(15))

Conspicuously omitted from the document were those few provisions favoring the residents, which were still in P.D. 705. For example, private rights were no longer given explicit recognition; even wildlife could no longer be hunted, trapped and harvested without prior ministerial approval. (Section 15) Permission had to be obtained, presumably also by members of cultural minorities, to enter and occupy watershed reservations and make kaingin (Section 10a) and to practise "upland agriculture." (Section 19) Former references to "open, continuous, adverse and public" occupation of land were dropped. The possibility was also raised of the closure of watershed reservations to prevent their "exploitation, occupancy, or development" whenever the Minister deems this to be necessary in the "public interest." (Section 4h) As before, "resettlement programs" were proposed if deemed "for the interest of the public." (Section 11)

* * * * *

Philippine legislation for forest lands and watershed areas in the 1970s reveals a major change in development perspectives. This followed increasing commitments by Philippine lawmakers to "accelerated" national development. The original administrative-regulatory concerns were increasingly supplanted by commitments to technically advanced ideas of organization and production as the two areas came to be related more consciously to national planning and development. With changes in the administrative and organizational focus, development thought turned in a technocratic direction. This change was quite rapid. It occurred within the relatively short span of ten years and seems to have been virtually finished before the advent of the 1980s.

The influence of technocratic values manifests itself in the objectives of this policy. These changed from a regulatory-administrative interest to large-scale utilization of forest and watershed resources for commercial purposes. They were related to the new thrust towards rapid economic development and, on the political side, to new ideas of political sovereignty based on the concept of the nation-state and the associated concept of "national interest." They were accompanied by the increasing tendency on the part of lawmakers to view development objectives in technical rather than conventional legal and administrative terms and to use scientific approaches and concepts for pursuing their objectives. Sweeping powers were given to the state over the planning and utilization of the country's natural resources. Private corporations were invited to contribute their technical knowledge and experience to the exploitation of resources.

A technocratic mode of thinking can be also seen in the transfer of watershed reservations to the jurisdiction of the Ministry of Energy in 1979. At the time of the transfer, this was a new government agency, as one Manila newspaper put it, "relatively new and vigorous,"[6] unencumbered by commitments made in the past. This enabled it to get away from the more conventional outlook of forest supervisors. Moreover, a ministry placed in charge of only one specific resource (energy) was likely to view its task in a more technical way; unlike the old agency, it did not need to bother much about the many administrative, physical and human problems that often arise when many interests are present.

Technocratic influence in this policy was perhaps most pronounced in the provisions of the codes having to do with occupancy or residency. These provisions tended to treat people as objects rather than subjects of development. The forest dwellers were now referred to as " primitive tribes" and " valuable assets," and were viewed purely from the perspective of effective resource management and planning.

It is evident that the policy-makers who formulated the legislative acts discussed here increasingly embraced concepts associated with technocratic values. Their ultimate goal was to maximize national production by using technical or technological means in the context of universal planning, which is the hallmark of technocratic thought. There were also immediate consequences of the new acts which should not be disregarded. The codes may be seen as attempts to divest the inhabitants of forests, particularly the cultural minorities, of their land or to formalize such divestiture. The rights of the minorities to " ancestral domains," although legally recognized since 1909, had been continuously assailed since Spanish times and the codes were but another instance of this trend.[7] The divestiture of such lands would make possible the application of more effective management and rationality in the context of planning and development. In this respect, the codes of the 1970s went beyond traditional attempts of the bureaucrats of the forestry department to assert their authority, based on the belief that forest land is a patrimony of the state and that only they have the right and capacity to maintain it. References to their traditional hostility to the minorities miss the technocratic thrust of the new policy. The speed at which the legislative acts were introduced and the thinking underlying them suggests an intervention of technocratic thought. As noted above, this is reflected in the fast transition from administrative-regulatory to technical approaches and the increased prevalence of advanced concepts of management science as well as in the ultimate objective, which was large-scale commercial exploitation of Philippine forests in the context of overall national planning and development.

In the next section we turn to two case studies to illustrate the themes raised in this chapter. The Cellophil

Resources logging operations in Abra and the attempt to undertake a massive hydro-electric project in the Chico Valley watershed provide a good illustration of the new technocratic legislative and political framework as it was applied to the control of natural resources and impacted on the lives of ordinary Filipinos.

Development Projects

The Cellophil Resources Corporation (CRC) and the Chico River Basin Project were both located in Northern Luzon.[8] The first project, established in 1973 and operated under the Herdis Management and Investment Corporation, was a wood and wood-processing enterprise. This was a huge concession for timber and pulpwood, covering some 197,000 hectares of land mostly in the province of Abra. The company was a highly mechanized " modern" type of corporation and was expected to greatly increase the capacity of the country for pulp products and foreign earnings through their export. The company's main plant was located at the confluence of Abra's main rivers so that rivers instead of roads could be used for transporting logs from the uplands and for waste disposal from the mill.

The company sought to justify its presence in the region both on rational economic grounds and in terms of its local benefits:

> The first streaks of industrialization and urbanization are dawning over the province of Abra. In a geographic area where a poor farmer has to wrestle with the harsh elements, coax the soil into fertility and implore Providence for a seson of grace, a new promise appears.
>
> In the erstwhile rural town of Tayum, people suddenly found themselves in the middle of a maelstorm of industrial activity. In the barrio of Gaddani particularly, carabaos that once tramped through

> ricefields with the languorous rhythm developed by the ages of rustic existence gave way to tractors and bulldozers levelling the land, trucks transporting personnel and material.[9]

The reaction of the people living in the areas affected by the CRC projects was at odds with the promise of this technology-based view of devlopment. In Abra these were largely members of the Tinggian cultural minority. Like the other cultural minorities in the region, they had lived in their territories since time immemorial, had a traditional mode of organization and practised mixed sedentary and rotation-cultivation economy based mainly on the plantation of irrigated wet rice, river fishing and raising livestock. The advent of the huge CRC project was a dramatic challenge to their economy and way of life.

An early manifestation of the Tinggian position was a resolution of March 4, 1977, by the residents of barangay Tubo. This expressed concern for some of their customary rights and values as a consequence of Cellophil's activities in their territory:

> . . . in the long run we will not only lose our right over the land and privileges on its resources but also our freedom and security will also be lost because of our close relationship with other municipalities and provinces by having a peace pact which our forefathers have been sacrificing will be destroyed due to the misunderstandings between those parties brought in by the company.

This criticism was soon followed by others of a similar nature[10] which accused the company of violating the Tinggian conception of property (especially communal ownership) and of seeking to manipulate and divide the people as well as of importing immoral habits (e.g. by encouraging beer gardens, gambling dens and so on).

A more comprehensive statement of the position of the Tinggian community was the resolution of an intertribal budong (peace pact) meeting held at Bangilo, Malibcong in September 1978. Unanimously approved and signed by 1,265 "Tinggian People," the document demanded "first priority to use of the lands, waters, and other natural resources which our ancestors have used, and defended in Abra since time immemorial" and asserted that "the best way to preserve our heritage of peace, cooperation and unity" is to give protection to their rights to communal forest, pasture land and private agricultural land "which are the sources of our livelihood." They criticized the Cellophil company for threatening "our very existence as a people" because it "has adopted a systematic policy of withholding vital information about its schemes and activities in the concession areas thereby generating fears of eviction and displacement among us."

A similar position was restated at the inter-provincial meeting held in January 1979 in Tiempo, Tubo. This led to the passing of the "Resolution of the United Cordillera" signed by 180 representatives and addressed to the Prime Minister and President Ferdinand E. Marcos. As before, Cellophil was assailed for seeking to destroy the people's livelihood and cultural identity. The company's entry, it was declared,

> will decimate the culture, traditions, customs and other such vital interest of our people . . . [and] will ruin the strong unity and harmony of our tribal communities which are held together by an extensive peace pact system . . . [also] will cause the moral, cultural, economic as well as physical displacement of people and the loss of their ancestral lands which they have safeguarded for centuries.

In the same document the company and the Bureau of Forest Development were accused of seeking "to harass us into submission" and members of the federation were prohibited from working with CRC or other companies without first consulting the peace-pact holders and the residents of the place.

Various tactics were applied by Cellophil to engender popular acquiescence to the company's presence in the territory. Attempts were made to persuade the people that Cellophil was "an industrial firm with a civic consciousness"[11] and a source of economic progress for local people. Another strategy was to replace incumbent officials with new ones sympathetic to the projects, as with the replacement of provincial Governor Balbin by Arturo Barbero, Abra's provincial development officer and the son of Deputy Minister of Defense Carmelo Barbero. There was an increased military presence which was justified by a rise in NPA activity.

Pressures were also exerted on local officials. In January 1979 a meeting of municipal officials of Malibcong at Bangued was held in the presence of Deputy Minister of Defense Carmelo Barbero to discuss the Cellophil issue. Officials were urged to accept an agreement, known as "Mutual Agreement," which would allow Cellophil to operate in the municipality under certain conditions. The agreement was signed on their behalf by Solomon Dayag, Municipal Mayor of Malibcong, but was subsequently denounced by many Tinggians despite the apparently generous provisions included in it. A similar occasion was the convening of a meeting in March 1979 by the provincial authorities in Camp Villamor, Bangued, which all elders and peace-pact leaders were ordered to attend. A declaration addressed to President Marcos was prepared beforehand, which some 250 participants subsequently signed. This denounced the Tiempo declaration, in which native leaders two months before had expressed their opposition to Cellophil's operations. It was stated that

> . . . we now . . . proclaim to the world that we are not forbidding the operation of Cellophil Resources Corporation [CRC] in Abra; on the contrary, we thank the Prime Minister Ferdinand E. Marcos for having established that company here in the Province of Abra so that we may also share its rightful benefits.

In addition, requests were made to the President to introduce a land classification program in all their towns, to obtain compensation from CRC for any damage done to their property, fields and irrigation systems, and to disallow the floating of cut logs down stream in towns where "the CRC has not yet begun operating or [which it] has not yet reached." The government institutionalized the new arrangement by establishing an "Association of Peace Pact Holders and Elders of Abra," which included a board of advisers headed by Governor Barbero and Colonel Lasanten, both Ilocanos, and eleven Tinggian municipal mayors. This arrangement was, however, not lasting. The elders, upon their return home, were reported to have rejected the Camp Villamor Declaration on the ground that it had been extracted under duress.[12] The reaction of the Tinggians to Cellophil was, as we will see, essentially the same as that of the Kalingas and Bontocs to the Chico River projects. In both cases huge industrial enterprises that were encouraged by legislation passed in the 1970s had an adverse impact on the people living in the affected areas and highlighted two different views of development goals.

* * * * *

The second project, the Chico River Basin Project, was a major power-generating enterprise. It was to comprise four sites for hydroelectric plants or dams with the total catchment area of 3,419 square kilometers, of which a large portion was to be submerged by the dams. Its total potential for power generation was estimated at 1,010 megawatts per year. Referred to as Chico I, II, III and IV, the first two sites were located in the Bontoc area of the old Mountain Province, and the other two in the territory of the Kalinga people. The project, on which the work started in the mid-1970s, was expected to inflict great damage on the physical ecology of the region and its social environment. It would involve the destruction of thousands of houses and flood fertile riceland. It would transform the existing pattern of the economy and have serious cultural and social implications for the people affected by it. But against these considerations, the project was to contribute greatly to national productivity and therefore to the Philippine economy. It was intended to help increase national self-sufficiency in electric power, which was a priority in national development, generate employment, and effect great savings in foreign exchange, as well as open up new areas for industrialization. It was also expected to speed up social and economic progress in the Cordillera provinces themselves, which were among the most underdeveloped parts of the country.

From the onset there were strenuously articulated differences in the respective views of officials and local people on the desirability of the project. Project officials, as might be expected, saw the project as modernizing and desirable, even necessary, for national development. This was not the position of the Kalinga and Bontoc people. They saw the project as threatening their way of life and cultural identity. Government spokesmen were unable to convince them that the new arrangement would allow them to continue enjoying their traditional interests and rights, including the use of their ancestral lands. There was a need for reconciling the idea of technological development with Igorot ideas of development. However, the persistence of controversy and conflict between project administrators and the natives indicated that this was never achieved and perhaps could not be achieved.

An early articulation of the Kalinga position on the project came from Bishop Claver, himself a Bontoc native of the Mountain Province. In a letter of April 25, 1975, addressed to President Marcos, Claver concentrated on the human impact of the project, identifying at least two basic concerns. The first was cultural; according to Claver, cultural concerns reflected the " deep and genuine feeling" the Igorots had for their ancestral lands and for their native beliefs and customs. He implied that the historical continuity of Kalinga culture should not be disrupted. The second concern was practical; it had to do with relocation and compensation. Claver expressed the fear of the natives that they would not be adequately compensated if they left their existing territories. He drew attention to the fact that the Igorots had a deep distrust of the government's actions and urged the President to have the government adopt a more open approach, adding that " the burden is on the government to explain itself to the governed."

In another document, published in 1975,[13] Claver observed that there was a conflict of values as to what kind of development was desirable, and the question to be raised was which values were to be favored and which sacrificed, a choice that, he said, involved "deeply human and ethical problems." He saw this as the basic dilemma underlying the Chico River project.

> The government seems to make economic development the primary value and is ready to use its coercive power to bring it about at any cost. The people directly affected by the project, on the other hand, evidently place more value on the intangibles of life: oneness with the past, with their forebears; the preservation of values of the spirit that spell out for them what is to be *people*. Which should get priority? It is conceivable, the world-wide need of cheap energy being what it is, that the Chico Project can be equated with national survival, and if so, then the right—even the life itself—of a small segment of the population must give way to the more sovereign rights of the rest of the nation. But can this equation be made honestly? And is the rest of the nation willing to be a party to the destruction of others of their kind in order for them to survive? This question must be put to the nation at large.
>
> In its simplest form, the question posed to all Filipinos is this: What price are we willing to pay for development—and development as defined in practice by the present government?[13]

A similar concern was expressed in a resolution by the clergy of the Vicariate of the Mountain Provinces at its meeting in Bontoc on September 29–30, 1975. This accepted the Chico River projects as potentially beneficial to the people living in the area, but assailed the manner "in which the government has tried to push these projects with almost total disregard for the people's rights such as to representation and participation in the decision-making process."[14] Also criticized was the government's inability to show any successful resettlement and rehabilitation scheme, and the lack of a serious dialogue between the government and the people, "compounded by the lack of credibility on the part of the government";

it was suggested that the latter was perhaps the main reason why people opposed the projects. In another passage, the position of the priests was summarized as " 100 percent for the dam but 200 percent for the people," and to stand with the people for no dam at this time, unless the position of the government and that of the people were reconciled.

On other occasions Igorot ideas of development were articulated by community leaders. A well-known instance of this was the Vochong Conference on Development held in Quezon City in May 1975 under the auspices of the Share and Care Apostolate for the Poor Settlers, with some 150 Bontoc and Kalinga leaders in attendance. The participants rejected the Chico projects and signed a six-point peace pact which specified various forms of obstruction to the projects. These included a prohibition on selling goods and giving food to project employees and workers, a strict prohibition against accepting bribes, and the like. Cordillera Igorots who were in the military were also warned " not to be hard" on those who opposed the dam.

By the early 1980s the problem of accommodating the ideas of development held by the mountain people remained unsolved. Indeed, Catholic Bishop William Brasseur, vicar apostolate of the Mountain Provinces, found it necessary to restate a point he had made five years earlier.[15] According to him, " We [the Church] are 100 percent for the construction of the dam but we are 200 percent against it, if it means the destruction of the ancestral land of the native residents without proper compensation, and if it means the violation of their human rights." A new proposal made by Brasseur was to introduce a 15- to 20-year moratorium to give the people more time to ponder on the advantages and disadvantages of the Chico River project.

Documents published during this period exhibit the strategies used by the advocates of modernization—project administrators or representatives of government—to gain popular acceptance of the Chico River projects. They relied on persuasion, manipulation, the use of forcible means and the like. Perhaps the best known was that associated with the government's intervention through PANAMIN (Presidential Assistant on National Minorities), headed by Secretary Manuel Elizalde. Supposedly designed to help the national cultural minorities "to uplift [them] socially, politically and economically," this government agency eventually became the main coordinating body for government action in areas affected by dam construction. However, it was increasingly condemned as behaving unethically. It was charged with deceiving and bribing people, using intimidation, manipulating the news coming from the region, trying to divide the Kalingas and showing disregard for Igorot cultural values.[16] The most condemnatory charge against PANAMIN came from the Catholic hierarchy in an open letter to the President in January 1977.[17] In this letter the agency's practices were described as "detrimental to the good of the cultural minorities and the image of government" and reference was made to such malpractices as "the coercion and buying of loyalty by PANAMIN officials, corruptions of the local leadership, forcing agreements and confessions, [and] sexual immoralities of a PANAMIN official." In a subsequent letter, again written by the bishops of the Philippines, PANAMIN was condemned for destroying the culture of the people. The bishops assailed the agency for attempting to foist a false notion of cultural identity on the cultural minorities.

A more co-optive approach was promoted by the government under the rubric of the Kalinga Special Development Region (KSDR), which was headed by the governor of Kalinga, T. Odiem. According to President Marcos, this scheme was aimed at improving the quality of life of the Kalinga people through socio-economic programs mobilizing government resources in encouraging self-help projects, providing improved facilities in education and fostering the development of Kalinga leaders.[18] Characteristically, the scheme's advocates tried to focus public attention on government priorities in development and the need for infrastructure to hasten the region's socio-economic growth. The new scheme, however, was confined only to areas affected by the Chico River project and so had a divisive effect on Cordillera society. It naturally aimed at sidetracking the issues that were of immediate concern to the people, like relocation and the accommodation of their traditional values.

Following the demise of PANAMIN the National Power Corporation (Napocor) took over the role of negotiating with the Igorots. Napocor's representatives tried to create a good relationship with them. The Napocor head, Gabriel Itchon, conceded the extent of Igorot resentment and talked about the need to restore confidence and credibility in the government's position,[19] blaming past "errors" on "an evident communication gap." However, his initiative was not successful. Napocor repudiated an earlier attempt at settlement, involving a quite generous compensation offer negotiated by PANAMIM (on the ground of excessive cost), which conceivably would have received support from the natives. After this the behavior of government agents became increasingly repressive. One reason for this was the growth of the NPA, by then widely supported in the Cordillera as a result of government intrusion.

These strategies throw light on the means government officials were prepared to adopt or tolerate to achieve their objectives. They reveal a lack of sensitivity to popular culture and a willingness to use violent means to suppress resistance. The lack of sensitivity on the part of government officials was an important cause of the increased hostility. This was a point made in 1979 by Bishop Brasseur in a letter to the Minister of Defense Juan Ponce Enrile,

> . . . if the national government agencies had used other means and another approach without violating the human respect due to these minorities, if previous promises made to the people who were in similar situation as, e.g., in Ambuklao and Pantabangan had been fulfilled, the government would have had more credibility with the people and, no doubt, there would have been less troubles . . . the people of Kalinga are not against the government as such. However, they resent the way they have been treated by government agencies . . .[20]

Bishop Brasseur implied that had government officials shown more openness and sincerity in dealing with people, some form of mutual agreement with them could have been achieved, which would accord with popular ideas of human dignity and development.

The persistence of such crudely manipulative and insensitive behavior among government officials draws attention to the presence of attitudes that reflect technocratic values. Obviously these officals held a modernizing, technology-oriented outlook, and being in positions of considerable authority, they were not prepared to compromise when the claims of technology and tradition were pressing in opposite directions. Their vision was fixed on the ultimate ends of actions, not so much on the means used to get there. The moral dilemmas arising from this situation were evident to some senior officials but they were only of minor importance. These officials were in a position to explore and recommend schemes that would cope with the moral issues present in the projects but they seemed to have largely ignored such issues.

* * * * *

These case studies point to the gap in perceptions of development between government officials and the people. The former believed in the beneficial forces of modernization, which they identified with technological progress. This was shown in the choice of the project, which reflected an emphasis on, in Richard Dorall's phrase, " technology, gigantism and integration into the ever-widening and complex national and international politico-socio-economic system."[21] Viewed from this perspective, the traditional Igorot mode of existence, unexposed to modern technology, was likely to be regarded as economically unproductive and socially backward, and tending to keep people poor.

Second, there was the issue of the popular reaction to the project. This was embodied in persistent opposition, which ranged from reservations about the project to complete rejection, in extreme cases expressing itself through support for the NPA. At the root of the opposition was a basic difference in attitudes towards development. As noted earlier, the Igorots sought to preserve their customary social arrangements which they saw were being threatened. Both the planning and implementation of projects were done in secrecy and people were kept guessing about the company's or government's next move. Only when the Igorots took the initiative and publicly expressed their dissent were government officials willing to consider their requests. Popular participation hardly came into the picture.

Third, there was the issue of the behavior of project officials and their strategies for dealing with popular demands. There was a lack of sensitivity to the values of the Igorot people, particularly when those values were in conflict with technological ideas of progress. This was reflected in the tendency to deal with people's problems in legalistic and technical terms and to disregard traditional authority structures and issues of culture and community needs, which officials were unable to handle. The conduct of public officials embodied elitist values based on the belief in the superiority of technology and technological ideas of progress and relative disregard for other values. There were a variety of stategies, used by officials to manipulate opinions, which only alienated further the native people from the government's position.

It should be said, however, that open hostility between officials and the local people was not necessarily present in their mutual contacts. In the Igorot community, modernizing values existed side by side with competing values. Still, to the advocates of technological change, appeals to traditional rights and beliefs were not likely to mean very much. Moreover, preference for technology was not necessarily the sole value present in determining the behavior of officials. There were also other, often competing, moral values which had to be accommodated and might not always fit comfortably with their technocratic outlook. The senior technocrats themselves would often voice concern for a humane treatment of the people affected by their projects.

It may be objected that insensitivity to people's demands or needs, elitism in conduct, and the manipulation of people by their leaders are not new but familiar features of both traditional authoritarian systems and modern bureaucracies. This point is not denied. However, in this case the technological-technocratic explanation appears to fit the situation illustrated by these case studies. First, before the adoption of the technological approaches in legislation, government administration was conceived essentially as a regulatory system. It was focused on the efficient discharge of regular administrative functions and followed the Weberian closed organizational system. There was no rapid or major departure from the established practice. Changes in its structure and behavior followed only the application of the new approaches and concepts, and were based on technical and professional standards and new economic concerns. Second, technocrats should not be viewed as ordinary bureaucrats. Because of their superior status as experts, they were relatively free in their choice of the means they used in development projects. The tendency on their part to persistently favor technological solutions suggests that they perceived activities in terms of technical problems that were quantitatively manageable and focused their interest on the means of actions, on how development was to be achieved, rather than on the end point of actions, the goal of development. Peoples' values and ideas of development were outside their field as experts, since they could not be easily objectified. Hence they tended to avoid such values or treated them as of dubious relevance. These two points suggest that the attitudes of government officials should not be viewed as merely reflecting an established authoritarian practice. Rather, they should be related to the particular context of change occurring in the Philippines in the 1970s, which suggests an emerging technological-technocratic outlook and practice.

Development Plans

The final section of this chapter deals with development planning, a topic of special importance in technocratic thought. Traditionally, technocrats have treated planning as the principal tool for introducing productive rationality and have emphasized the role of experts who have the necessary technical skills and know-how to maximize the use of physical and human resources. The Philippine development plans from the late 1960s to the early 1980s will be examined for the presence of technocratic concepts and values.

An overview of the Philippine development plans during this period reveals a gradual rather than a sudden departure from conventional approaches to national planning, focused on fiscal and monetary controls and limited government economic initiatives. The first national development plan of the late 1960s was the *Four-Year Economic Program for the Philippines, Fiscal Years 1967–1970.*[22] This was essentially a conservative document. Like similar plans in the past, this program was primarily an economic statement, outlining " the aggregate targets which the economy will seek to achieve."(p.7) In its economic policy, it followed established practice by dealing with such topics as fiscal and monetary operations, tax reform and the stability of exchange rates. Its estimate of the growth of the national economy was rather moderate, at about 2 1/2 percent annual increase in real income per capita. However, a fairly strong commitment was made, at least in its policy guidelines, to economic and administrative rationality. Proposals were made to cultivate " positive thinking based on self-discipline in work, in business, and in social and political life" in the implementation of the plan, and in administration it was recommended to overhaul the government machinery for development " to make possible the increasing use of economic rationality, as distinguished from social and ward-politics considerations, in the planning and implementation of the public investment program."(p.15)

The last objective is significant; it reflects the thought of Rafael M. Salas, who prepared the plan and who, as noted earlier, advocated a separation of economic or development planning functions from political functions, viewing the former as more important for national development.

The emphasis on economic rationality was reflected in the preference expressed for industrial development, which anticipated later economic directions. The document stated that one of its basic policies was "to hasten the change in the composition of manufacturing output toward basic and intermediate products."(pp. 32–33) On the other hand, the plan was conservative in its social orientation. It followed the usual budgetary allocation in this sector. About 30 percent of the total budget was allotted to social development and 83 percent of this was to be spent on education, mostly the construction of schools and the salaries of teachers. There was little money left to support other social services such as health, labor, social welfare and community development.

The Four-Year Development Plan 1971–74,[23] prepared by Placido Mapa Jr., and described by President Marcos in his Message as "the second phase of the integrated and comprehensive development program of the Philippines," did not depart significantly from the general orientation of the preceding document. It did, however, suggest some advances towards a broader concept of social change and greater economic and administrative rationality. Citing increased GNP as its basic target, the plan identified three main objectives: maximum economic growth consistent with stability (which includes "maintaining the past growth of the Philippine economy"), a more equitable distribution of income and wealth, and minimizing of unemployment.(p. 10) Going further, the plan made references to other important goals such as the development and enlargement of the middle class, the determination to "prevent unwarranted monopolies" and the commitment to develop "the necessary manpower for development." (p. 11) The plan also reflected a new awareness of "administrative requirements for development." It introduced a section which defined the planning processes required for plan implementation. Although social equity was stated to be one of the plan's principal objectives, the concept was not seriously elaborated.

The revised *Four-Year Development Plan 1972–75*[24] prepared by Gerardo Sicat appeared soon after the preceding plan. The justification given for the revision of the original plan was that it was desirable to move projections for economic growth upwards because of better prospects. The new plan introduced several significant innovations. It sounded a new tone in dealing with problems of national development and reflected more manipulative ideas of national planning.

First, economic development (defined as "sustained national prosperity") was again the principal objective, but this was conceived in terms of more dynamic economic growth and long-term planning objectives. For example, higher projections were made for GNP and it was claimed that the earlier development targets "have been fairly conservative and had, as a consequence, generated a psychology of conservatism in business expansions."[25] Emphasis was also given to the long-term effects of development objectives, which were now expanded from three to five: higher per capita income, greater employment, more equitable income distribution, internal stability and regional industrial development.

Second, although the main priorities remained the same as in the previous plan (land reform, manpower training, infrastructure, community development and family planning), ten new ones were added. These included housing, rural electrification, cottage industries, cooperatives and tourism, expanding the interest of the state to new areas of development and incurring more comprehensive government planning responsibilites.

The third innovation centered on the re-examination of the framework for economic policy. This involved an attempt to liberalize the economy by removing protection from inefficient industries and thus increasing competition. The significance of this proposal must be viewed in the light of the former protectionist practice. The new agenda for economic reform included Sicat's favorite idea of the maximum use of the labor force. The plan declared:

"A recurring theme is that increasing the degree of economic freedom and the reliance on the market mechanism in the allocation of products and resources is desirable. The necessity of exploiting more fully the economy's comparative advantage in labor-intensive industries will also be underscored."(p. 32)

Another major innovation was the introduction of a macro-economic framework for development planning. It was argued that this would assist in investment decisions by helping to identify the relationship between the relevant empirical variables and priority areas of the economy (e.g. employment and gross national product, money supply and prices, personal consumption expenditures and prices) and to target such variables to achieve particular growth rates. It would also assist in decisions about alternative policies or policy trade-offs by making planners "aware of the implications of the assumption of alternative policies, and, accordingly, of what targets are feasible under different assumed conditions."[26] Other innovations included an emphasis on government reorganization and regional planning and development, and a new section entitled "human resources development," meant to underscore the importance of manpower in planning, which comprised activities traditionally grouped under "social development" such as education, housing and nutrition. In the Annex, policy implications, implementation and monitoring were given emphasis. The plan stated that "the government is well aware that attainment of its development objectives . . . is heavily dependent on the nature and effectiveness of its economic and social policies . . . [and] satisfactory implementation . . . [and that this] calls for close and continuing attention to policy requirements," and added to this that "all public policy having development implications are being kept under constant scrutiny by the government." (p. 9) The concept of social justice was also mentioned. According to the plan, there was "the need to combine the highest possible rate of material growth with active promotion of social justice." The concept of social justice, however, was not discussed.

The plan was further elaborated in the State of the Nation Message of 1972,[27] which declared that it was designed for social change and that such social policies as land reform, land distribution, food production campaign and general welfare projects were justified because they would help " to eliminate the prime source of social discontent." (p. 25) Emphasis was also given to the economic strategy for development; this involved a massive capital program, which Marcos promised " as soon as possible to shift most of our expenditures to the capital requirements of growth, and make this the Administration's principal task." (p. 33) Dynamic action was promised in government administration, aimed at making the civil service " an effective instrument of development." (p. 83). This was to include a massive training program to make government employees " more efficient, more perceptive and more knowledgeable of the development goals of the government." (Ibid.)

The *Four-Year Development Plan 1974–77*[28] reflects the same combination of a dynamic developmental outlook with a technocratic methodology. A new element was a commitment to a wider distribution of economic benefits by more extensive social development and by integrating the various aspects of regional development. This was stated in the introduction by Sicat, who again prepared this document: " The Plan provides for a wider distribution of the benefits of economic growth by placing greater emphasis on social development and by integrating the approach to regional development." The social orientation of the plan was strengthened by the reinstatement of social development as a major section in the plan; this replaced human resources development, while manpower development was treated as only one aspect of social development, together with other aspects such as population, health and nutrition, education and social welfare.

As on previous occasions, proposals advanced in the plan were aimed at rapid economic expansion and reflected an interest in rational and comprehensive approaches to development. The macro-economic framework was again given prominence, and relevant targets were again specified. A variety of New Society reforms were cited to highlight the wide scope of recent policy achievements. There were new programs mostly aimed at generating industrial expansion in such areas as manufacturing for export, agri-based and resources-based industries and engineering industries, including shipbuilding. Emphasis was also given to industrial dispersal and the expansion of infrastructure, including the construction of new superhighways and the extension of rail networks. A similar emphasis was given to the tourism sector. The Appendix of the Plan highlighted the problems of energy generation by including a section entitled " National Energy Plan and Policies."

Two other strategies were given prominence. One was the integrated regional development program, which sought to coordinate local or regional development with overall national development. The justification given for this was the need to spread the benefits of economic progress more evenly and to attain a more balanced spatial distribution of development activities throughout the country, and ultimately, greater social equity. Another strategy, which aimed at establishing an effective machinery for development planning, included proposals for the identification of priorities for project development and implementation, for improvement of the regional data and information system to support planning activities, and similar concerns.

The last plan belonging to this period is the *Five-Year Philippine Development Plan 1978–1982* (including the *Ten-Year Development Plan, 1978–1987*).[29] This was perhaps the most technically advanced document in the Philippines up to the early 1980s. More than before, it reflected a preference for economic and administrative rationality and comprehensive planning. Although it made an explicit commitment to humanist ideas of development, as a working document for dealing with development activities, it suggested a basic commitment to science. The same impatience to get quick results which characterized the two previous plans underlies this document as well.

The humanist commitment is found mainly in the introductory part of the document. The plan stated that it was "dedicated to the pursuit of advancement of total human welfare. It is an instrument towards overcoming poverty, underemployment and unemployment," adding that "the achievement of a much improved quality of life for every Filipino is the supreme national aspiration."(1-1) In another passage, the plan said it was "aimed primarily at rectifying grave economic and social inequalities in the course of our ascent to nationhood" and to embody "our concerted attack on the problems of mass poverty, unemployment and underemployment, and social justice."(XV) This humanist theme was also explicitly developed in the statement of "national development goals," of which three out of seven aimed at the improvement of the human condition: promotion of social development and social justice, increased development of lagging regions especially rural areas; and improvement of habitat through development of human settlements and proper management of the environment. (1-6)

The concept that was given central attention, however, was comprehensive planning. This was presented as an instrument for turning "our vision of the future" into reality. The justification given for the "primacy of planning" (xxix) was the need for "coordinated and effective management of development." (xxx) The earlier efforts at national planning were assailed as greatly inadequate because they had been "hindered by poor organization, bureaucratic rivalries, and the old society habits," but it is contended that the planning machinery had now been "rationalized and strengthened . . . in order to harmonize the work of all agencies involved in development efforts, as well as to make individual activities mutually reinforcing at all levels of implementation." (xxx) An illustration of such emphasis on comprehensive planning is in the treatment of "social development," which was the concept revived in the preceding plan for describing social planning. This concept was now dropped in favor of the "social sectors," suggesting an attempt at integrating social with economic planning.

As in the previous plans, the main preoccupation was with economic growth. A number of objectives and strategies were explored in this connection, including the proposals aimed at intensifying the process of industrialization, particularly the manufacturing sector, which was expected by the year 1987 to be equal in importance to agriculture. (xxxv) In the area of economic integration, attention was given to the "Integrated Area Development Approach"; this was adopted for the improvement of the countryside to "ensure the interlinking of factors critical to the development of the depressed areas: infrastructures, farm-to-market roads, rural electrification, institutional innovations and arrangements, research and development, credit, marketing, and social services, among others." (xxxviii) Similar strategies were proposed for other important development projects and aimed at dramatic increases in production through dispersal-expansion-integration. They included such schemes as Integrated Cattle Development, Animal Dispersal, Expanded Fish Production, National Cotton Development and Cooperatives Development. Efforts at rapid economic growth were reflected in the proposal to establish new major industrial enterprises, including an integrated iron and steel mill in Misamis Oriental; ammonia-urea, nuclear power and petrochemical plants in Bataan; a Kraft long-fiber pulp plant in Abra; various ship-building and repair facilities; and an export processing zone in Mactan, Cebu.

Another topic given prominence was development administration, which was viewed as the most effective method for implementating development policies. Its principal task was defined thus: "to support the achievement of national goals and objectives." It was promised that during the decade of 1978–1987 support would be given to "a continuing program for increasing administrative capability through dynamic flexibility and self-reliance."(p. 411). This included a number of schemes for increasing the rationality of administrative systems. One of those given prominence was increased regional autonomy, which was referred to as "further decentralization of national government operations and the evolution-ary development of regional government within the context of enhancing local autonomy."(p. 411). (In one regional plan, this scheme was described as "the main strategy for development of the national government" and as involving "the decentralization of operations to the regions culminating in the institutionalization of regional governments."[30]) Closer reading of the text, however, suggests that what the plan had in mind was not necessarily devolution of administrative or political authority but, rather, more effective central control over local planning and development. The scheme seems to have been aimed at getting local authorities to participate more actively in national planning, and so help to mobilize and better control local resources for national development. This was reflected in a passage in which regional governments were referred to as primarily "agents of the national government in its development efforts." Similarly, in another passage, it was stated that plans, programs and projects of regional governments should "concur with those of the national government and its agencies."(p. 412).

Another scheme meant to increase administrative rationality centered on planning integration. It was proposed that regional plans should be integrated with national plans to tighten up "existing vertical and horizontal relationships of planning entities" and that planning should be oriented "towards synchronizing and harmonizing the sectoral and spatial dimension of the national and regional plans." (p. 413) A number of similar integration schemes were introduced in other parts of the government system. They had a variety of objectives. Some aimed at creating "a competent workforce" and proposed "an integrated career and personnel development program . . . to effectively support national development efforts" (p. 416) to be brought about by "an intensified training program" for government personnel at all levels. Others aimed at reorganizing the government machinery for area development planning, for instance, the Bicol River Basin Program, or the development of criteria for the assessment of government corporate structures "based on both legal and functional considerations."

(pp. 418–419) On the same integrative principles were based a number of other schemes for creating an effective evaluation system for budgetary programs and projects, improved management methods, improved auditing methods, upgrading statistical capacities of government agencies, developing evaluation of coordinative mechanisms for the formulation of plans, a more effective system of communication and information dissemination, and the institutionalization of effective monitoring and control systems.

This overview of Philippine development plans between the late 1960s and early 1980s reveals a trend in national planning towards technology and science-based ideas of development and the increasing authority of technocratic thought in Philippine policy and administration. In the early years the technological-technocratic element in government planning was relatively weak. Government planning was limited in scope and was concerned mainly with the regulation of routine matters of public interest. There was, however, a gradual departure from the conventional practice manifested in an increasing emphasis on economic and administrative rationality for accelerating and diversifying industrial growth.

The technological-technocratic orientation was already strongly present in Gerardo Sicat's first plan. Policy objectives were expanded to cover a major portion of the economy, long-range planning was introduced together with a macro-economic framework as an important feature of national planning, and liberalization of the economy away from protectionism was proposed. A commitment was also made to extensive government reorganization, which included the administration of regional development and planning, and to similar reforms for strengthening the economy and making government systems more responsive to national development objectives. Social development and social justice were given relatively low priority.

Sicat's second plan strengthened the new commitments and policies. Rapid expansion of the economy was again the plan's primary concern and the means to achieve this was to be comprehensive planning. Innovations concentrated on big industrial enterprises, export orientation, integrated organizational approaches, including integrated regional development and planning, as well as creation of an effective machinery for development planning. Although recognized as important, social development was treated, like the economic sector, as one of the sectors of planning characterized by economic problems to be resolved by the proper use of available resources.

The central concept of the last plan was "the primacy of planning." Apart from comprehensive planning, preference was shown for the rationalization of economic and administrative systems in the context of development, large-scale industrial enterprises, centrality of national over regional or local planning, and technical reforms for increasing administrative and planning capabilities of the government system. Among other approaches highlighted in this document were advanced concepts and techniques of technology and science; these included systemic and integrative approaches and advanced O & M theory and practice. Although the plan declared a commitment to humanist ideas of development, it tended to treat social problems as essentially economic in character, that is, problems accessible to action through government economic planning. Significantly, the concept of "social development," revived in the preceding plan, was abandoned in favor of the "social sectors," with its more technical and manipulative connotations.

There was, then, a changing perspective in development thought in Philippine government planning in the 1970s and early 1980s. The influence of technological-technocratic concepts and values in national planning had grown to encompass a major portion of the national economy. The plans, however, also made strong commitments to humanist objectives of development, which in the later plans were declared to belong to preferred national goals. But while these commitments were mentioned in the general statement of development policies, they were not clearly reflected in the policies themselves. There was no major shift in the plan towards the social sectors. For instance, in the last plan only 14 percent of the funds for the Philippine "infrastructure investment program" for the 1978–1987 period was allocated for the social sectors, reflecting a low national priority, for these sectors included a variety of basic public services such as health, nutrition, family planning, education, manpower, social services and community development. (*D.P. 1987–82*, Table 2.7, p. 30) The commitments in these areas were mostly translated into the essentially technocratic language of development planning and economic development, and in the process were robbed of their original humanist character.

Findings

Government development policy in the Philippines, as it is reflected in legislation, development projects and development planning, have been canvassed in this discussion of development thought. The discussion reveals a fairly consistent value pattern in the direction of technology and science-based ideas of development, which are at the heart of technocratic thought. The dominant thought had changed in a remarkably short period of less than a decade from conventional, regulatory ideas of government functions to aggressive, dynamic ideas of development characterized by the advocacy of advanced methods of technology and science. This is brought out with particular clarity in the studies of legislation and national planning. By the end of the decade and in the early 1980s the authority of technocratic thought had spread widely and penetrated deeply into the whole Philippine government system.

The commitment to technocratic ideas of development is reflected, first, in changing development goals with the new emphasis on rapid modernization and industrialization through technological means. There was an associated displacement of humanist, people-centered values by management-of-resources-oriented values alongside the progressive abrogation or diminution of traditional rights in many areas. This was particularly manifest in the behavior of officials towards the indigenous people affected by such infrastructural projects as the Chico dam construction and Cellophil concessions. Commitments to more " humanist" concepts of development were not entirely lacking; however, they were given a low priority. They were confined mostly to general statements of national goals and did not seem to have significantly affected actual policy-making.

The strong commitment to technocracy is expressed perhaps most explicitly through the strategies, concepts and approaches that were to be used in these policy areas. These included advanced concepts of management and organization, concepts of economic and administrative rationality, rationalization of systems, centralization of decision-making, strategies for increasing administrative and planning capabilities, and long-range and macro planning. Prominence was also given to integrative models, the idea of liberalization from economic protectionism and to comprehensive national planning (" the supremacy of planning") which embraced the management of all available Philippine resources, physical and human.

On the behavioral side, tendencies suggesting a technocratic outlook permeate government documents and statements on the projects. Officials tended to view problems in terms of technology and planning and were unresponsive and insensitive to traditional ideas that they perceived as obstructing their planning. They had an elitist cast of mind, and the result was a general lack of communication and consultation, accompanied by the use of threats and force to overcome resistance to project implementation.

CHAPTER V

The Roots of Technocratic Thought

The prevalence of the technocratic mode of thought needs to be explained. Three factors in particular have played a crucial role in its emergence: a commitment to science, support by political leaders, and a reinforcing educational and recruitment policy. These appear to have been the "necessary conditions" for the emergence of technocracy, but considered by themselves they were not "sufficient conditions." Although other factors might have been present, these three are the ones that keep reappearing.

Commitment to Technology and Science

A commitment by the Philippine government to technology and science was the central factor impelling development ideas in a technocratic direction. This followed a pattern of choices by leading officials and politicians in favor of methods, techniques and concepts that originated in science and were meant to inject dynamism into the process of national development. It involved a gradual and cumulative rather than a rapid acceptance of scientific approaches; the logic of this commitment had taken more than a decade to become fully manifest. This commitment was reflected in the political rhetoric of the time, particularly that of President Marcos, and in the behavior of officials noted earlier. It had several interrelated aspects. One was a commitment to the model of modernization advocated by advanced industrial countries. This had been a persistent theme in Philippine development, affirmed by President Marcos: "we must modernize . . . if our nation is to continue and prosper."[1]

Another aspect was the advocacy of advanced technology. This was elaborated by President Marcos in his *Five Years of the New Society.*[2] Marcos considered two types of technology as relevant to Philippine conditions. One was the technology appropriate for labor-intensive types of development. He contended that this sort of technology is good for developing countries like the Philippines, where labor is cheap and capital is scarce. The second was advanced technology and capital-intensive approaches. According to Marcos, this is the technology necessary for rapid industrial growth, with the labor-intensive alternative playing only a secondary role. Indeed, given increased commitments to modern methods of industrial production, there may not be much choice but to follow such high-technology strategies. Moreover, the substitution of labor for capital is not always possible. Marcos explained:

> But for all this predilection for labor-intensive technology in the Third World, there are those development planners who favor capital-intensive technology, because it is believed that this type of technology produces "a distribution of incomes favorable to profits, and in turn, to capital accumulation." Moreover, it is argued that "the very attractiveness of industrialization stems from its promise to bring modern techniques to a backward economy and to embody them in power and machines," and for the most part machinery used in the developing country will have to be imported from the advanced countries. In a certain sense, the basic processes of industrial production are rather rigidly determined by existing technologies in the advanced countries, and labor intensity can only be injected by the developing country in the indirect and supplementary processes of production, such as handling, packaging, and distribution.

> Likewise, the factors of capital and labor are not as readily substitutable for one another as is often supposed. There is a general deficiency of skills in the developing country that makes for technological constraints on the industrial production process.[3]

Marcos's technological prescription for Philippine modernization was a mix of the two types. Ultimately, however, with his preference for "a rapid drive towards industrialization," which he called "the overriding objective of development,"[4] he came out in favor of advanced capital-intensive technology.

The government's commitment to technology and science was reflected in an increasing preoccupation with professionalism in the public service. This was, of course, not new but now it had been given a new emphasis and related to national development goals. A high regard for professionals in public employment was expressed in 1973 by Jose Crisol, Undersecretary of Home Defense. According to him, "The professionals in this country should be the leaders in improving and modernizing its social, political and economic institutions," and, he added, "it is the professionals who inject quality to community life."[5] Imelda Marcos, the First Lady, glowingly described the professionals of the country as in "the forefront of the New Society," and expressed her "unlimited faith in their capacity and their expertise, their learning and their programs." According to her, "they deserve as much honor as war heroes or great statesmen" and must be "emulated."[6] Government statements referred to the institutionalization of professionalism in the civil service through the establishment of training programs meant to upgrade professional qualifications and to enhance requirements of professional competence.

Closely connected with the emphasis on professionalism was the recognition of the special status of technocrats in national development. This is exemplified in the lavish praise accorded them by President Marcos; he called them "indispensable in the management of change" and seemed to view technocracy as the highest stage of bureaucracy.[7] In another passage, Marcos described the graduates of DAP (Development Academy of the Philippines), a breeding ground for government technocracy, as "the new breed of government executives" and suggested that professionals like them "hold the key to the attainment of the goals and values of the New Society."[8] Elsewhere he referred to them as the "democratic elite in the public service"[9] boasting that his administration had brought into the government more technocrats than any of his predecessors and remarking that "most of the members [of the cabinet] were pirated from the private sector."[10] In sum, technocrats were for Marcos a body of persons of excellence, a non-political elite who were to replace the "political vanguard" of traditional political thought. Aided by a supportive political leadership, they would be instrumental in transforming the entire Philippine society in accordance with the goals of national development.

Finally, the commitment to technology and science assumed a strong ideological dimension. This was because "development" became the center of the government's concept of progress and because the process of development came to be seen, to a large extent, in scientific and technological terms. The latter seems to have encouraged a symbiotic relationship between development and science; thus "science" became part of the official ideology. This particular relationship was recognized by Marcos in a short essay on Filipino ideology, published in 1979 and entitled simply *Towards a Filipino Ideology*,[11] where he made specific reference to "scientific ideology." While the essay retained the traditional idea of the "New Filipino" as the goal of social transformation, it departed from the usual formula of ideological model-constructions. Its focus was not merely on the goals of development, but also on the "scientific" strategies or means for achieving such goals, as in the advocacy of a non-dogmatic and open outlook in dealing with development issues. "If an ideology is to be scientific," Marcos stated, "it must necessarily be an open ideology, or partial. It must allow room for modification and growth, and thus, cope with novel contingencies."[12] Here Marcos linked his ideology to the need for flexibility and adaptability, two influential concepts of organization and management science. Another illustration is to be found in Marcos's treatment of development goals. These included self-reliance in basic needs, a high level of industrialization, rural and regional development, and resource adequacy and development, expressed in operational terms, which made them manageable for use in the planning and implementation of government policies and projects.

Yet another illustration of this orientation is the emphasis on comprehensive planning. The justification given for planning was that in order to get desired results we should make decisions today or "plan ahead" and that there was a need to "lay the foundations for the coordinated and effective management of development."[13] This implied that without comprehensive planning the desired objectives of development could not be realized. In sum, the commitment of the Philippine government to technology and science had a number of related dimensions, including the adoption of advanced modernization theories and techniques, increased professionalism, the recognition of a special role of technocrats in national policies and development, as well as attempts to incorporate the scientific approach into the official ideology of Marcos's regime. These had a mutually reinforcing effect and made the "scientific" element in Philippine development thought very strong.

Support by Political Leaders

The promotion of technocratic ideas by political leaders was another crucial factor contributing to its increasing influence. This support manifested itself at three levels. First, in the choice of methods and concepts for dealing with policy problems which carried the stamp of technocratic reasoning and planning. This point has been elaborated in the preceding section. Second, in the use of organizational strategies favored by technocratic planners, such as centralization and rationalization. These strategies were endorsed by Marcos's policy-makers. Third, this support was reflected in the adoption by Marcos of an ideology that made an explicit commitment to technology and science.

The growing tendency towards centralization and rationalization of the government system has been described in an earlier chapter. This generated a "new dynamism" in development and was symbiotically linked to the creation of an authoritarian system alongside a system that was conceived essentially as a mechanism for integrating and mobilizing human and natural resources for achieving the goals of national development. Under martial law the country's political structures were part of a vast bureaucratized and centralized politico-administrative system in which technocrats played a prominent role as central administrators. This pattern was followed at national and sub-national level. It was observed both in barangays and local governments, which, despite the statements of national leaders in support of enhanced local autonomy, were used to mobilize local people for national development and often to promote ideological conformity. The direct government supervision of local government activity was vastly intensified.

A similar trend toward rationalization and centralization of systems was also discernible in government policy-making. This was suggested in a 1977 study of de Guzman et al., which concluded, unsurprisingly, that ordinary people were not part of substantive decision-making, whether in initiating or implementing policies, for "the technocrats assumed as given their rationality and correctness in preparing the decrees to be transformed into policies."[14] The study identified a tendency to displace conventional rationality, based on freely operating political structures, by technocratic rationality.

Government statements about political and administrative centralization and rationalization must, however, be qualified. They reflected commitments or ideal objectives, but not necessarily actual conditions. There are many reasons why such commitments were not very effective in practice. Alongside the residual parochialism of the pre-martial law period there were substantial political, linguistic and other differences among the provinces which were hard to overcome. Government regulations were unworkable for a variety of reasons including lack of local sources of funding, lack of effective supervision and poor enforcement when they ran up against traditional opposition. Thus the Local Government Code passed under Marcos was applied only in a limited way.

The technocratic concept of "de-politicization" also contributed to changes in Philippine development attitudes under Marcos. It involved the view that decision-making could be done in a technical way rather than through traditional political means. It suggested the possibility of leaving to technical experts policy issues that in liberal systems were subject to political debates. Technical approaches were now to be applied, not merely to the management of national economy, but also more broadly to public management and political systems.

The technological orientation in Marcos's ideology was partly described in the preceding section. It was identified with his purported scientific interest in non-dogmatic, open approaches, systems adaptability, the "Human Settlements" approaches, the use of advanced organization and management concepts, industrialization, the management and development of natural resources, comprehensive planning, and so on. Marcos's ideology also embraced political principles like the notion of a "positive and active government." His overall ideological position encouraged government intervention and comprehensive planning, and reflected the technocratic idea of expert-managed systems. It drew attention to the ends rather than the means of development, allowing its advocates to adopt large-scale methods of manipulation by the state to speed up the process of development.

Supportive Educational Policy

The national educational policy developed in the 1970s and early 1980s was another important factor in the emergence of technocratic thought in the Philippines. The wider diffusion of technocratic values could hardly have occurred without its support. Government statements during this period stressed that education was to be related to the government's effort at accelerating the country's social and economic development. Two concepts were given particular emphasis: relevance and practicality. They kept reappearing in the public statements of national leaders (as in President Marcos's observation that education today has a "need for relevance" and that it should reflect "the realities of national life") and in periodic revisions of the school curriculum. Marcos referred frequently to "education for development . . . [bringing] national development into the center of instruction and learning."[15] Relevance was also highlighted in the policy objectives outlined in 1981 by the Ministry of Education and Culture. The document talked about "refashioning of curricula to eliminate useless subjects and introduce what the new education bill decrees to be 'relevant and innovative' courses to align the schools with the developmental programs of government," and defined its aims as "to develop new curricula and systems to align teaching with the developmental programs of President Marcos."[16] Elsewhere there were references to "a period of redirection and adjustment in the Philippine educational system," and its stated objective was to make educational priorities consistent with national development priorities. Government documents also identified several revisions of the teaching curriculum that had been introduced to meet the need for such relevance. They included a variety of new teaching subjects such as population control, family planning, ecology, tourism, conservation of natural resources and similar topics.

Practicality was closely related to relevance. Viewed from this perspective, education sought to create an attitude of "work-orientedness and production" or "work orientation and manpower development," and, according to one government publication, encouraged young people through incentives "to take up technical professions to ensure an adequate supply of skilled workers in a world where technology dictates the lifestyle."[17] The expected growing demand for trained manpower and professionalism led to reforms of the school curriculum in a technical direction, which included new programs for manpower-training, establishment and upgrading of technical institutions, non-formal training programs for out-of-school youth and the unemployed. They also included specific projects like regional manpower-training centers, science teaching centers, agricultural-skills training projects, and counselling and guidance services[18] as well as the establishment in 1981 of a bureau for vocational and technological education whose stated objective was "to push through the desire of President Marcos to strengthen this aspect and align the educational system to the manpower and economic needs of the country."[19] In sum, the thrust of martial law educational reforms was in a technical and vocational direction. But these reforms also seem to have been motivated by political considerations. The emphasis on technical and vocational subjects de-emphasized the liberal arts, where students and practitioners had been among Marcos's most vocal opponents.

These changes in the teaching curriculum were not, however, universally adopted. They were slow to take hold in higher-level institutions. This is attributable in part to the fact that most Philippine colleges are private institutions and are outside the sphere of direct government control, and in part to vested interests established in teaching, which could not be easily overcome. Moreover when one considers the millions of scholars filling Philippine tertiary institutions every year, extensive reforms would have been very costly and could have led to unmanageable demands on manpower and equipment. More importantly, there was little pressure from the public for change. One reason for this lay in the fact that Filipinos were mesmerized by college certificates, regardless of content. There is something ritualistic about this obsession, perhaps because it is rooted in traditional concerns for maintaining social status and "appearance." Thus the modernization of the educational system lagged behind the government's national development rhetoric and its commitment to a more professional and more "practical and relevant" national policy of education.

More decisive results favoring the technocratic outlook came from the continuation of overseas programs of study which started in the late 1950s and from developments in post-graduate studies in the Philippines itself. Under the overseas programs, gifted scholars were selected for specialized studies in American universities on Ford Foundation or similar study grants. Domestically, new institutions were set up which were expected to provide professional education at specialized post-graduate levels. These included the College of Public Administration of the University of the Philippines and other specialized institutes at the same university, the Development Academy of the Philippines and the Asian Institute of Management. These institutions shared a commitment to productive efficiency, ideas of scientific management, related economic concepts and professionalism. They closely followed American curricula and adopted the dominant methodology of positivist social science. The influence of this elitist system of education, proud of its "professionalism," was not only confined to teaching the principles of government administration and economics but to some extent permeated, through these elite institutions, the whole Philippine tertiary educational system. Other colleges or institutions followed the prestigious institutions, particularly the U.P., in their curriculum and teaching practice and often hired graduates from such institutions for their teaching staff. Furthermore, the influence of the American foundations-funded scholars (some belonging to the so-called "Harvard mafia") was considerable; many of them became government ministers under Marcos. Obviously, this system of education was likely to facilitate and perpetuate technocratic ideas for managing organizational systems in both the public sector and the private sector. In addition, contemporary studies[20] indicated that technical and professional background was becoming an increasingly important factor in government employment, partly because of the mushrooming of government agencies dealing with technical matters and national planning.

These three factors—a commitment to technology and science, a supportive political system, and supportive educational and recruitment policy—had mutually reinforcing effects in conditioning people strongly exposed to them towards a technocratic way of thought. The conditions favoring technocracy were further fostered by the absence of alternative conditions normally present in liberal democracies.

CHAPTER VI

Technocratic Influence and Marcos

The findings of this work show the growing influence of technocracy in Philippine policy-making. This was illustrated earlier through the reforms introduced in the Philippine government system and in the areas of Philippine legislation, development projects and national planning. These show commitments to rapid modernization based on technology-based ideas of managed development, preference for comprehensive planning, systems rationalization, integrative approaches and advanced organization and management concepts. " Humanist" issues of development were not seriously considered or were treated as aspects of economic planning.

Technocratic Influence

The pervasiveness of technocratic influence is hard to determine on the basis of this work alone. However, government administrators were likely to be a great deal more development- and technocracy-minded in the 1980s than they had been in the early 1970s. (A 1974 study of the Philippine higher civil service showed only average " development orientedness" and concluded that " only a small fraction (16.6%) of the higher civil servants are highly committed to development."[1]) The reason for this was the expanding recruitment of experts and professionals and the persistent exposure of administrators to technical and technological ideas adopted in government systems, which was reflected in the increased use of technological-technocratic concepts and means in the planning and implementation of policies and projects.

There are at least two other reasons why the values of Philippine administrators moved in a technocratic direction. One was the influence of NEDA. Virtually all government systems found themselves caught in the web of universal planning done by this powerful central planning agency. This meant that nearly everyone in government administration was compelled to participate in the NEDA game of development planning, which permeated the entire government system; it included higher administrators who, as the 1974 study of the higher civil service indicated, contributed to planning less out of conviction than out of their inability to avoid such a responsibility. The second reason for a technocratic orientation was a persistent appeal by President Marcos to the country's administrators to adopt technological approaches in the civil service and government systems. Such appeals were expressed in the rhetoric of national survival or national development and were aimed at motivating administrators to accelerate national production. It is impossible to say to what extent the influence of NEDA and Marcos's appeals helped to inculcate technocratic values among Philippine administrators. Nonetheless these values became a part of the structural arrangement of the government system and could not be disregarded.

The strong influence of technocrats in the government system is also reflected in the role which the leading Philippine planners, together with the World Bank and the International Monetary Fund, played in the Philippine economy. This became particularly pronounced in the late 1970s and early 1980s during economic debates between the "transnationalists" and the "nationalists." The former position, held by the leading technocrats, advocated liberalization of import restrictions, lowering protective tariffs and promotion of export-oriented industrialization. The nationalist position advocated import substitution, protection of Philippine industries and often used the nationalist argument to criticize the government's economic policies. It had taken several years for the transnationalists led by three influential ministries—Economic Planning (Sicat), Industry (Roberto Ongpin) and

Finance (Virata)—to change the course of Philippine economic policy in a transnational direction. The rising power of Philippine technocratic planners has been well documented, as well as their close cooperation with the World Bank and the International Monetary Fund, the two international lending development agencies.[2] There was apparent congruence in outlook between the officials of the two agencies and the leading Philippine technocrats. The support of the two organizations for the transnational direction of the economy seems to have been crucial in having such policies adopted. These organizations had enormous financial resources which were needed for the country's economic growth or in times of financial crises, and therefore had the power to influence the direction of national development. Technocratic influence was further strengthened when Cesar Virata, a technocrat, became prime minister. This was reflected in the adoption of the transnationalist-World Bank idea of structural adjustments of the economy, which included restructuring of industrial and financial policies, as well as a more interventionist role for the state in bringing about such changes.

The rising technocratic influence is also documented in a number of policy studies of development projects. They highlight the presence of technocratic values and the lack of humanist values, and frequently ask the question: "Development for whom?"[3] One such study considered critically the relationship between government organizations for rural development and rural development as such. The author put the blame for frequent failures of rural development projects on the technocratic values of officials who implemented such projects. He argued that officials tended to impose their technocratic standards upon the system which did not fit the given conditions and made rural projects unworkable in practice:

> Despite the great attention devoted by the Administration . . . to restructuring rural organization and participation in development schemes, early results suggest that form rather than function represents the actual outcome. The top-down technocratic character

> of the new arrangements seems too complex for the average rural person to comprehend, or perhaps, more important, too alien to his own sense of appropriateness of how land and people really interrelate. They do not assure him the security he desperately seeks before he becomes open to experimentation.[4]

He added that "the model of the Filipino farmer that technocrat-planners carry in their heads does not conform to the reality out there."[5]

The technocratic element was also identified and criticized in a study of people's participation in community organizations. The study concluded that "the realization that where community participation in matters affecting people's lives is poor, the difficulty lies not so much in the people themselves . . . but rather in the constraints on action imposed by bureaucractic structures and restrictive institutional arrangements."[6] The writer suggested that this was because rural organizations were dominated and manipulated by "external" forces. Two such forces reflected the familiar technocratic approach: transplantation of alien advanced development concepts, principles and procedures that did not fit local economic and socio-cultural conditions, and rigid government supervision and control.[7]

Marcos and the Technocrats

The extent of technocratic influence can be determined with more precision when it is seen against the role played by President Marcos in policy-making. This question is not easy to answer. It poses particular difficulties for writers on dictatorial systems, for they have to decide whether the history of a country under such systems (like Nazi Germany) should be explained in terms of one dominant person (Hitler) or of a network of actors and events, including individuals, groups or party interests, often pulling in different directions in an effort to increase their power and dominance.[8] A similar question can be asked about the Marcos martial law regime, namely whether the behavior of the Philippine system was shaped by Marcos alone or by a variety of actors and events.

Political analysts treat the Marcos authoritarian era basically from two perspectives. The first highlights the dominant role of Marcos's personality in national policies. Marcos is portrayed as an able politician who, having attained the country's highest office, was eager to ensure his political survival and continuation in power. He is seen as the prime mover of national policies, while government officials were merely acting under his orders. He is said to have enjoyed absolute powers by virtue of his lawmaking powers and because countervailing powers were lacking. The whole thrust of martial law policies and reforms was toward centralization under Marcos as the supreme patron, which made his actions a blend of modernizing and traditional practices. The declaration of martial law is interpreted as essentially an attempt by Marcos to extend the tenure of his presidency, which was then about to expire. Marcos's good relations with technocratic planners is explained by their usefulness to him. However, he was never one of them, and could have done without them if he had wished to do so. His refusal to accommodate the wishes of the mountain people in the Chico River Project is explained as a matter of political and economic expediency on an issue on which Marcos apparently believed most Filipinos would support him. He seems to have been convinced that this problem would be resolved by political means and did not anticipate the deterioration in the situation that occurred later. On this view, too, the ideology of the New Society is regarded as essentially false and hypocritical, as an attempt to boost the attractiveness of Marcos's regime and legitimize his position. The ideology did not take firm root, and its rapid disappearance in the wake of the demise of the Marcos regime seems to lend support to this allegation. Similarly, the preoccupation with "human needs" and "humanist" ideas of development, carried out with such missionary zeal by the First Lady, led to superficial results and did not outlast the regime.

The demise of the regime is explained mainly by Marcos's failure as a leader and as a person. This is attributed particularly to his inability to contain the economic crisis of the early 1980s and to certain structural weaknesses in the government system and the economy. According to Sicat[9] too much autonomy was given to some agencies, and this militated against effective integration and coordination and led to fragmentation of the administrative and decision-making process; in addition, the growth of government monopolies allowed by Marcos was undermining public commitment to a market economy. Sicat also highlighted the extravagance and excessive self-confidence displayed by Philippine leaders following the early economically successful years of this regime. This led to undisciplined fiscal policies and extravagant public spending (he called the Ministry of Human Settlements "the conglomeration of the most dispensable public corporations in the present government"[10]) which lasted until the end of the Marcos regime. Among the factors that contributed to the demise of the Marcos regime was Marcos's corrupt practice of distributing public benefits to his "crony" associates and the looting of the country by his family. Marcos's cronies enjoyed access to government patronage and protection for their conglomerates during financial crises. In a relatively short time they came to dominate a substantial portion of the national economy, and so, according to Sa-Hyung Park, "threatened not only the administrative autonomy of the technocrats but also the material base in the private sector of other co-opted elites,"[11] who saw them as competing with them for economic power and influence. This led to a split among the governing elites, with some of them now openly criticizing aspects of the Marcos regime. In the words of Lela Garner Noble, "Party switching was replaced by defections; policy-making, once subject to executive-legislative stalemate, was now hampered by bureaucratic rivalries."[12] From this perspective, which centers on Marcos's political survival and power, the role of technocracts was seen as secondary to Marcos's political interests.

This view of Marcos's role in Philippine policy-making can be contrasted with one which highlights the basic rationality of policy decisions under Marcos and gives strong recognition to the role of technocrats in policy-making. It is emphasized that Marcos had nothing in common with the "irrational" dictators of the fascist variety, who demand fanatic devotion and use emotional symbols or force to compel people to follow their message. Rather, he always insisted on the legality of his regime and on proper constitutional procedures, and took care to have his reforms legitimized by laws or "presidential decrees." A case can at least be made that Marcos's major policies can be explained by a basic commitment to the modernization of the country. The values underlying his legislative reforms and development plans in the 1970s reflect technocratic approaches that were used to speed up the modernization process. Both Marcos's ideology and ideas on social development can be viewed in the modernization context. His ideology not only encouraged national growth but also drew attention to the desirability of using concepts and methods of technology and science. In retrospect this ideology was vilified as merely another instrument used by Marcos to legitimize and strengthen his position, but this disregards the fact that its main elements had already been defined before the martial law regime. Furthermore, this was not the way the New Society ideology was viewed in the early part of the Marcos regime. Its later demise was caused by the administration's failure to enforce it through the educational system, as well as by the Filipinos' lack of discipline and failure to follow the ideology once its novelty had worn out. It was also undermined by the bad example set by the First Family, particularly the First Lady, who indulged in ostentatious displays of wealth, and by the regime's increasing reputation for corruption. Similarly, social development, which incorporated an interest in meeting basic human needs and improving the quality of life, increased in scope and was subject to rational planning. This might have been concealed from the public by the flamboyant and personal style of Imelda Marcos, although behind the facade of her person there was a solid core of bright young technocrats of the Ministry of Human Settlements who orchestrated such ideas and the accompanying schemes that were to put them in practice. The failure of these schemes seems to have been due, to a large extent, to the planning component that went into them, which, as noted earlier, tended to deny people's needs rather than to help meet them.

In this policy-making context the technocrats were increasingly relied on in the making of public decisions. As noted above, their influence became particularly strong in the early 1980s, when the debate between the "nationalists" and "transnationalists" over the future direction of the economy (which followed similar debates under Macapagal and earlier administrations) was won by the transnationalists with the support of the World Bank. A triple alliance was forged based on the transnationalist faction comprising the government, industry and finance, with decision-making safely anchored in the hands of transnationalist technocrats.[13] Indeed, in mid-1981 Marcos announced: "I am going to sit back and let the technocrats run things."[14]

This profile of Marcos's leadership highlights the basic consistency of Marcos's position and the strong role played by technocrats in Philippine policy-making. However, it shows that the policy-making process was constantly undermined by economic and administrative practices that lacked rationality in the context of technocratic planning, and resulted in fragmented decision-making, extravagant spending and lack of fiscal discipline. It also draws attention to the role of Marcos's cronyism in the downfall of his regime. Two factors, in particular, undermined confidence in his system. One was the gradual decline of the economy and the economic crisis of the early 1980s, which rapidly eroded popular belief in the ability of the regime to improve the lives of ordinary people. The other factor was political. This included rising political unrest, a growing opposition at home and abroad (e.g., the pressure of the US Government to return to "democracy"), increased NPA activity, army and police violence, and an increasing breakdown of law and order and lack of public discipline, because the authority lines in the regime were increasingly dispersed, and political "moro-moro," or make-believe, was ineffective in situations that called for decisive leadership and action. This resulted in a loss of confidence, especially amongst the commercial class, in the ability of the leadership to provide security and stability. The decisive event alienating ordinary people from the regime in power was the assassination of Senator Benigno Aquino in 1983, for which the regime itself was put under suspicion of complicity.

At this later stage of the Marcos regime, events were getting beyond Marcos's control and his political and economic options became more limited. The ultimate inability of the regime to act is illustrated in the crony system. The practice started during a period of economic boom, with the grant of a few economically useful concessions to loyal friends; but with increased self-confidence and extravagance these concessions were extended, eventually growing into a system, and Marcos himself got caught in a network of obligations of his own making, which damaged his reputation and limited his ability to act. A similar paralysis characterized the New Society ideology at its later stage. By then it had lost its inspirational drive and Marcos did not have other ideas to put in its place. His promises of a return to "democracy" were too vague and could hardly be trusted. It is interesting that during this period, national technocrats like Sicat sought to dissociate themselves from many of Marcos's practices. A survey of their opinions conducted in 1984 showed that most of them regarded "crony capitalism" as the main cause of the economic crisis and favored a political change with "free and credible elections" and "formalization of the procedure for presidential succession."[15] However, they cannot be exonerated from the events leading to this crisis, for it was their advice that Marcos followed in major economic policy formations. The crisis of the early 1980s occurred at a time when the planners, in collaboration with the World Bank, had introduced a new direction for Philippine economic policies. This was towards export-oriented industrialization and was introduced at a time of growing international protectionism and debt crisis, which was very unfavorable for new export ventures. Thus Philippine technocrats bear considerable responsibility for the poor state of the economy then and the crisis of confidence accompanying it.

This picture of technocrats' and Marcos's role in Philippine policy-making indicates that technocratic influence was extensive, but on the whole less dominant than it has sometimes been alleged; as Mark Turner put it, "the technocrats have not attained and consolidated the preeminence in economic policy claimed by Bello."[16] Two reasons stand out which undermined technocratic rationality and made major changes in the system difficult, if not impossible. One, noted above, was the continued presence of political patronage with the associated problems of structural weaknesses in the political system, extravagant government spending and undisciplined fiscal policies. This conduct greatly weakened the technocrats' position, which now, as Lela Noble put it, "seemed limited to developing plans, which were not necessarily implemented, and to negotiating the outside funding required to subsidize an increasingly expensive system."[17] The second reason was the limitation imposed on technocratic intervention by the existing structural arrangement of the economy. This was based on private enterprise and was moving towards increased economic centralization and involved growth of new monopolies. Economic power and wealth were concentrated in the hands of a relatively small number of business groups and these were allowed to operate with little government interference. John Doherty in his study "Who Controls the Philippine Economy?" came out with eighty-one families spread among ten banking groups who controlled the economy.[18] This imposed considerable limits on the possibility of exercising technocratic rationality in Philippine planning and policy-making.

CHAPTER VII

Technocracy, Its Impact and Reorientation

The adverse impact of technocracy on popular conceptions of development has been described throughout this work. This chapter seeks to explain why this state of affairs exists. It examines the nature of the technocratic position to show that the ministrations of the technocratic elite are not people-oriented and have an adverse effect on human freedom. It also explores the possibility of overcoming or reducing the associated dualism of values as well as the possibility of a reorientation in technocracy's development.

The Nature of Technocracy

The technocratic position described in this work has a number of characteristics. First, it is based, as we have seen, on technological ideas of progress. Limitations on progress are not set by prevailing spiritual values but rather by the physical constraints imposed by existing technology and science. They are technological rather than legal and have to do with the way in which people utilize the things they find around them. Social reality only reflects the more basic technological arrangements. Works like the *Study Course of Technocracy* argued that a rational social and economic order can be brought about only when the outlook of society changes in a technological direction, and this is to be done by a systematic application of the principles of technology and science. The association between technocracy and technology can also be established through the link of the two to the idea of modernization. Technocracy followed the rise to prominence of modernization ideas, which highlighted the crucial importance of technological change for social progress. Indeed, technocratic thought may be viewed as a product of particular theories of modernization.

Second, the technocratic position follows the conventional methodology of social science. This is derived from positivist science and draws a sharp distinction between facts and values. Only the former have objective validity while the latter have merely subjective reality; scientific activity is viewed as "value free." This methodological approach makes comprehensible some aspects of technocratic behavior. It helps to explain the relative disregard by Philippine technocratic planners of social and cultural values, which they viewed as being outside the area of scientific or technocratic intervention and not relevant to their judgment as experts. This also explains why they excluded values from their policy design. Despite their commitment to scientific objectivity and value-free science, however, the technocratic position is ambiguous. This is because the technocrats' actions are based on certain assumptions which suggest some basic value commitments, of which we will have more to say shortly.

Third, there is a particular assumption about human nature underlying the technocratic position. This is that human nature is malleable. It can be shaped and reshaped to suit the form that is wanted. The transformation of people and society becomes a matter of manipulating structures to achieve the desired results. Some such assumption seems to underlie President Marcos's determination in the 1970s to change the Philippine system through a series of institutional reforms. It was given explicit recognition in 1974 by Minister Blas Ople, a technocrat. Invoking the authority of B.F. Skinner's theory of positive reinforcements, Ople contended that "The spectacular rise in various indices of development since the proclamation of martial law—especially in peace and order, economic production and social and civic discipline—demonstrate that sweeping changes in environment can result in sweeping changes in human behavior." Pursuing this theme further, Ople added that "The results confirm the behaviorists: an enhanced environment will produce enhanced human behavior . . . environmental changes will induce desirable social traits."[1]

This view of human nature is embodied in much of the official behavior described in this work. In the *Study Course* the learning process referred to Pavlov's experiments in reflex conditioning and to theories of stimulus and response, and their mechanical operation was given prominence. It was also stated that "all human beings respond alike to a like external environment," and man was described as "an engine taking potential energy in the form of chemical combinations contained in food, and converting this potential energy into heat, work, and body tissues."[2] It may be noted that, in this view, changes in social norms and mores are relatively easy to achieve.

Fourth, there are implicit ideological assumptions derived from the notions of scientific rationality underlying this position. The "goals" aimed at in the technocratic vision are the rational organization and management of social life, which is taken to be a good in itself and to reflect the classic idea of the Good Life. The ambiguity of this position is reflected in the intimate connection between the goals and the means used to achieve them. The goals are conceptualized in terms of big, comprehensive models, macro thinking and systems approaches. Traditional ideological goals are not necessarily rejected, but they are treated as relevant only to the extent to which they fit into a science-based ideology of development. This explains why technocrats often profess different traditional ideologies and why their behavior is unlikely to be monolithic.

The emphasis on the bigness and rationality of systems gives prominence to the need for dealing with issues in relation to the objectives of the whole system rather than the means used to achieve such objectives. This often encourages a willingness among technocratic planners to sacrifice the interest of people today for a "better life" in the future, and to ignore the negative impact of development projects on minority groups if they embody advantages for national development. Another criticism of the emphasis on bigness is that this disregards or considers lightly the actual conditions of life and that it stimulates the practice of making people fit into technocratic models rather than the articulation of models arrived at through popular decisions.

Fifth, a technocratic preoccupation with techniques based on technology and science is another characteristic of this position. This involves considerable danger for human development. According to Hannah Arendt, the uses of such techniques have serious adverse and dehumanizing effects when applied to people. She refers to them as secret weapons for enforcing organizational or bureaucratic conformity, for they imply the idea of deviation from the mean. They tend to treat people as objects, not subjects of development. This was exactly the criticism raised in the Chico River project, that "the Kalingas find revolting the manner in which the progress of civilization is being imposed on them [through technical manipulation]. They resent being treated as objects and commodities."[3]

Sixth, the technocratic position is characterized by the notion of systems rationalization, which encompasses three dimensions—economic, political and organizational. Systems rationalization refers to attempts at improving administrative systems in order to increase their predictability and stability and to maximize their efficiency. In the economic area, this comprises a variety of strategies such as monetary and budgetary policy controls, forecasting and comprehensive planning, liberalization of the economy as well as regulations meant to encourage industrial production and productivity. These strategies dominated Philippine economic thinking in the 1970s and early 1980s. Systems rationalization as it applies to politics involves the replacement of liberal approaches by "technical" and "disciplined" problem-solving approaches, based on decision-making theory and science. There is also a tendency to displace traditional political and ideological thought by what is sometimes called the "end of ideology" state of political development, where, as Susanne Bodenheimer pointed out, "the ideological politics . . . can be replaced by a 'rational,' consensus-directed politics in which technical solutions may be devised for hitherto political problems."[4] Such a "rationalization" of politics was present in the Philippines during the martial law regime. It is reflected in Marcos's attempts to de-politicize public life and in his highly centralized form of decision-making.

In the organizational aspect, the concept of technocratic rationalization gives prominence to the role of organizations in social systems and development. This notion is of central importance. We noted that early Philippine advocates of technocracy like Sixto Roxas viewed problems of development in essentially organizational terms, as having to do with planning, management and coordination. Indeed, the technocratic revolution is essentially an organizational revolution, for it goes further than Max Weber's notion of administrative rationality, defined as efficiency and economy. It includes the broader concept of integration of functions in the context of comprehensive planning and development as the following quotation indicates:

> This [organizational rationality of technocracy] is . . . taken to mean pushing more vigorously such Weberian ideas as behavioral universality and impersonality in administrative conduct . . . [It also means] extending the boundaries of Weber's idea of coordination of efforts (in Weber's conservative account depending largely on the hierarchic structure of organizations, where authority is strictly delegated from above) to more dynamic, more encompassing conceptions of organization, symbolized by the ever-present word "integration."[5]

Despite obvious similarities in their emphasis on organizational efficiency and technical aspects, there are significant differences between the conventional (Weberian) and the technocratic type of organization in their respective goals, structures and operations.[6] The Weberian bureaucracy is essentially a "closed" system, inward-looking, focused on what happens within the system. The ideal technocratic organization is an "open" system, related to the broader context of national development. The former is structurally rigid, with tasks strictly defined, the latter is ostensibly adaptable and flexible; the former lays emphasis on the stability of systems—their routinization, institutionalization, etc.—the latter on the need for institutional change in a modernizing direction; the former professes value neutrality, the latter, while professing the same, tends to commit itself to "development" as a desirable value; finally, the former views officials as mere "advisers" (on the principle of the separation of policy-making/politics and administration), the latter treats technocratic administrators as experts and allows them a more positive role and more discretion in the policy-making process. Our earlier discussion indicates that the Philippine government system was strongly committed to the notion of organizational rationality. This is attested to by the ever-present references to the need for "integration" of systems and for comprehensive national planning. Indeed, all three dimensions of technocratic rationality were present in Philippine development planning during the period discussed.

The last prominent feature of the technocratic position is technocratic planning. This follows the pattern characterizing other technocratic concepts and activities. It is highly technical, done "from above" by experts, employs macro models, rational approaches and aggregate figures, and is quantitative and comprehensive in scope, covering the main goals of national development. It is identified with a high degree of professional competence and a mastery of advanced planning techniques. This approach to planning is said to have many advantages over conventional approaches, as the following comments of experts indicate:

> The specific knowledge of planning, which particularly constitutes the value of planning as a means for the preparing for decisions . . . differentiates planning from the advice in the traditional sense . . .The accumulated knowledge of planning is different from that of the traditional advisors in so far as that they . . . possess administrative-technical and legal disposition knowledge . . . while the qualifications of the new, often younger expert who does not possess

such knowledge, stem from the command of the planning techniques.

> Because of its undisputable conceptual superiority, comprehensive planning alone is in the position to satisfy the intellectual perfectionism often to be met in the planner and his demand for solutions which are theoretically exact . . . The macro-problems concerning the economy as a whole are often self-evident and seldom offer material for political controversies; the concrete individual projects, on the other hand, often conceal a great deal of political explosive.[7]

In this approach to planning, macro models are the crucial element and are associated with the favorite technocratic concepts of rationality and integration. The macro focus allows a broad rational outlook, but it may distort reality and lead to undesirable practices. It often generates development programs and projects by "planning alone," that is through planning models that are to be applied (or imposed) rather than through the painstaking path of micro research, and so tends to neglect project planning and divert attention from actual needs and policy implementation. One illustration of this was the practice of doing regional planning through collecting figures nationwide at the central planning office (NEDA) and then disaggregating them for regional uses. The earlier criticism of technocratic rationality also applies to technocratic planning, that this tends to increase state control, to overcentralize systems and ignore local interests. This was also the Philippine experience with national planning. It is reflected in the confident statements of national planners that they "design our future" and "consciously plan" and about "the primacy of planning," as well as in the technical concepts and methods used in this area.

Technocracy and Freedom

The present analysis shows the considerable limitations of the technocratic position, particularly with regard to people-oriented development. Another limitation stems from technocracy's impact on freedom. Preoccupation with freedom—freedom to initiate, freedom to choose—is one of the basic values of modern man. Yet government systems exposed to technocratic influence manage people from above. In the Philippines this behavior was exemplified by the greatly expanded administrative control of the state, the conduct of officials that suggested elitism and often arrogance, and the domination of national interest over local and individual interests. This situation, in which the claims of a managed state tend to come into conflict with those of freedom, draws attention to the danger to freedom from technocracy-dominated systems. It calls for an examination of the implications of increased administrative power for human freedom and of the place of freedom in such systems.

It may be argued that so much power in the hands of the state exercised through administrative means leads to increased state control over people and reduction of their freedom. There are economic implications as well, since this lessens discretion and sets considerable limitations upon personal initiative. Indeed, this condition evokes a picture of the Administrative State, in which people are both served and completely controlled by the institutions of the state. This type of state has been defined as follows:

> The essence of the administrative state is that men are born and live not merely . . . in subordination to power, but under a regime in which they are both served and controlled by an institutional complex composed of organizations.
>
> The distinguishing features of this administrative state include the delegation of responsibility for exercise of these various service and regulatory tasks to a

variety of specialized administrative agencies and officials who operate, not necessarily in accordance with fixed rules and customary standards of right and wrong, but rather with a measure of discretion—freedom of choice—that is sometimes unpredictable.[8]

Examples of such managed systems are many. One modern totalitarian system is Stalinist Russia, where all aspects of human life, political, social, economic and cultural, were controlled by an omnipotent state, and private initiative was repressed. A similar outlook is reflected in the various schemes for ideal societies such as James Burnham's technocracy-based oligarchy, in which political power is in the hands of technocratic managers, the old capitalist class is eliminated, and workers play only a servile role by contributing to production. A famous example of this idea in literature is George Orwell's book *Nineteen Eighty-Four*, where " Big Brother is watching you" always, seeking to control people not only physically, but also intellectually and emotionally. But the same outlook may be present in different degrees in real systems in which state institutions move in a technocratic direction. It was reflected in the Philippines during the Marcos authoritarian regime, and more broadly it is present, perhaps in more insidious ways, in modern industrial systems which are dominated by large corporations and in which public power tends to be subservient to their economic interests. Galbraith writes about the " close fusion of the industrial system with the state"[9] and shows how the two systems manage or manipulate society through planning in the interest of the industrial system. He urges that " the monopoly of the industrial system on social purpose be broken."[10]

The danger of unrestrained technocracy for democratic societies and human development seems to be real. One question is whether the threat to freedom posed by technocracy can be prevented or reduced. There are several reasons for believing that this may not be easy. To begin with the ideological climate of the age appears to be sympathetic to technocratic ideas. This point has been highlighted by sociologists like Mannheim, who have discovered a tendency in modern social systems to turn political and ideological problems into administrative and organizational ones. When operationalized, the original problems tend to be approached and solved in technical terms.[11] The growth of technocratic influence stems from the technocrats' mastery of technical expertise, particularly planning, and their ability to command advanced models and techniques of social science. Vente suggests that the politician who lacks such gifts often feels at a great disadvantage and " withdraws under the fascination of the technocratic model,"[12] leaving decisions, particularly in complicated and technical policy matters, in the hands of technocratic experts. He may even press them to provide " clear answers" and possible " solutions" to policy problems. Such a mastery of techniques also has an important social impact. Mannheim has pointed out[13] that improved social techniques not only possess greater effectiveness but that such effectiveness promotes domination by a minority. Thus technocratic experts become part of a dominant minority, a " class" composed of technocratic administrators and managers together with their corporate and bureaucratic masters.

It is evident that technocratic power can be formidable. Given favorable conditions, technocrats can control and dominate organizations and people through organizations. Despite such obstacles, there are, nonetheless, some possibilities for containing threats of technocratic domination. In liberal systems, for instance, the principle of countervailing powers can be applied or its application can be strengthened; on this principle there are multiple sources of political power and a variety of social interests that tend to check and balance one another. However, to be effective, this would require a vigorously operating party system, a free press and strong public opinion, which would allow the public to discuss openly and oppose publicly what they believe to be undesirable or abusive actions of technocratic administrators. Equally important, if not more so, may be the role of educational, scientific and intellectual leadership (Galbraith calls them " the educational and scientific estate, with its allies in the larger intellectual community"[14]), which, according to him, should become alert to the danger of technological domination for freedom and " need to assert the other dimension of life."[15]

Keeping technocracy under control, may not be easy, however, for another, deeper reason. This was identifed by Max Weber as a problem facing modern bureaucratic systems. This is that political and technocratic interests are basically in conflict. Democratic politics is concerned with individual freedom and accommodation of people's demands and self-interest, while technocracy concentrates on functional rationality and efficiency. In this conflict, as the preceding discussion shows, technocratic experts have many advantages. They have made themselves indispensable and have created a momentum of their own which neither autocratic rulers nor democratic politicians can afford to disregard. Given their strong position, it is an open question as to whether technocratic power will assert itself as the dominant social force. With respect to bureaucracy, Weber seems to think that the process of bureaucratization cannot be halted; it is part and parcel of the legal-rational order and values that are the dominant force of the age. However, bureaucracy can be treated as an integral part of political systems, as one of their sub-systems. Thus to keep bureaucracy under control, it is not necessary to dismantle bureaucratic systems; rather, it requires balancing their interests and claims to power with other interests and claims. Weber's argument is applicable to technocracy. This could be treated as one competing interest among many in an open political system, however strong its voice, based on the authority of technology and science, might be.

Reorientation of Technocracy

One basic issue that requires further examination is how to frustrate counter-humanist tendencies of technocracy-oriented systems and their threat to human freedom. Some possibilities offer themselves. One is to modify such systems by instilling more liberal values through reviving the traditional teaching curriculum. However, this move might not be very effective if technocratic values were to remain dominant, since it could lead to a confusion in values. It would involve only (what Unger has called) " partial criticism" of technocracy but might be insufficient to change basic social beliefs and practice. Another possibility is to keep technocracy in check by institutional means and enhance its public accountability and so reduce its threat to freedom. This follows Max Weber's prescription for the control of excessive bureaucracy. Technocratic power is viewed as representing one among many competing interests and its influence is balanced by other countervailing powers. This approach would require a " participatory" mode of decision-making, which would prevent technocratic planners from making decisions on important policy issues " from above" with limited accountability. It would call for decentralization of power, which would help to break the monopoly of technocratic decision-making at the center and would strengthen the interest in locally-oriented development, aimed at meeting village and regional needs and using appropriate technology.

The mere presence of countervailing power may not, however, be enough because technocratic power over politics and society tends to increase. One reason for this, noted earlier, is the growing dependence of policy makers on technocratic experts for " advice" because of the complexity and technicality which underpins modern policy-making. Under President Corazon Aquino, we shall see later, this influence was magnified because of her pragmatic, " non-ideological" management style. Another reason is that technocracy partly derives its power from certain "insidious" methods of control used in modern organizations. Blau and Schoenherr define three such methods which go beyond the usual bureaucratic control through the hierarchy of authority, rules and regulations, traditional incentives and the requirements of the machine. They are: control through expert power, through selective recruitment and through the allocation of resources.[16] The first of these methods is particularly important in the propagation of technocratic control; it consists of appeals to workers' "professional" commitment and aims at establishing a work environment that appears normal, "neutral" and rational because it is based on requirements of technology and science. In this case the use of money or physical force is not required to achieve compliance. Exposed to such methods, people tend to accept the technocratic position uncritically.

Another possible way to deal with problems created for freedom and development by the presence of technocratic values is to subject technocratic administrators to direct political control. In my study, technocracy has been treated as a relatively autonomous system. It enjoys considerable discretion in decision-making and in many respects can be distinguished from ordinary bureaucratic values and practices. This makes it difficult to subject it to normal political control and also to restore a government service more responsive to calls for a "better quality of life" and humane policies. For this reason it may be desirable that instead of seeking to increase the "responsiveness" of technocratic systems we should be seeking to destroy them. Arguably, bureaucratic systems, because they would be under direct political control, would not create the problem for freedom and human development which technocracy does.

There are, however, several reasons for believing that abolishing technocracy would be ineffective in the developing world. First, the need for expert knowledge and "consultancies" is recognized in all developing countries and their leaders depend heavily on technocrats to provide it. These conditions are unlikely to change in the near future. Second, keeping technocracy within the bounds of traditional bureaucracy seems to be possible only in systems with a long tradition of bureaucratic discipline and assertive political practice, e.g., in France, the U.K. and the U.S. In the new states these conditions are lacking. Third, it is unlikely that the present trend toward privatization or "smaller government" will change the situation. With continued population growth in developing countries and the magnitude of social and other problems, rational approaches through planning may become even more important than in the past, although the institutional form of planning may change in a more decentralized direction.

Two other proposals are given in this book, both seeking to encourage new approaches to problems created by technocratic values. One of these aims to strengthen the capacity for political leadership and initiative by educating politicians in the art of policy-making and analysis. This suggestion came from Yehezkel Dror, according to whom politicians should be encouraged to upgrade their qualifications by attending special courses "where they will spend between four and twelve months undergoing a very extensive experience designed to improve their performance as political leaders."[17] Such an experience would make them more competent to review policy proposals submitted by technocratic planners and to assess the merits of such proposals in the broader social and political context. Dror's proposal could be modified to suit particular needs or contexts.

The principal proposal of this book is to "reform" the concept of technocracy by changing its methodological perspective. It is argued that the counter-humanist attributes of technocracy can be traced to the conventional methodology of social science adopted by technocracy. At root is a rigid separation of facts and values, which leads to a denial of an objective status to values. This does not necessarily mean that technocratic planners deny the presence of human values in development activities; rather, they tend to ignore them in practice. They view them as "external factors," as mere background information, outside their professional interest, and hence leave them out of their project designs. Indeed, values are often treated as so many constraints on actions and perceived as obstacles to accelerated development. They create particular problems in liberal systems in which individual preferences are not only tolerated but tend to be encouraged and viewed as beneficial, for such preferences are often in conflict with the claims of technocratic rationality.

This proposal aims to overcome the mentioned separation of facts and values. It is contended that this cannot be overcome unless both facts and values are included in technocratic judgement. It is not enough that technocratic administrators accept as desirable certain commitments to holistic development objectives; rather, they should make sure that such objectives are incorporated into their design for policy formulation and implementation. This approach envisages, then, the integration of ideas of human development with technocratic programs of action. It has been explained as follows:

> If real development is to be pursued, not mere economic growth . . . it is imperative to incorporate somehow the ideal objectives into the working model of development administration. Without such action, the stated ideals are likely to be treated as more or less external to the model, as imposition on the model, hence unlikely to be integrated into actual administrative actions, save . . . as mere constraining factors on such actions.[18]

The new approach is likely to have important implications for social study and research. In the first place, it will broaden the scope of scientific inquiry. It will encompass under the wing of social science many aspects of social life that have been omitted from it in the past. It is not rigid and closed, as the old method based on " hard" scientific data has tended to be, but is open to a variety of factors, including moral values, holistic ideas of progress and a liberal belief in freedom. It should make not only technical but also value concerns objectively manageable. Elsewhere I have described this approach as:

> . . . a more flexible, more "utilitarian" methodological approach . . .—[characterized by] less concern for a strictly scientific outlook—[which is necessary] if the method of science is to liberate itself from some of its former limitations, such as those due to its positivist legacy, and to relate itself meaningfully to new types of experience.[19]

Second, this approach will have a considerable impact on the making and evaluation of development policies and projects. Unlike before, these cannot now be evaluated merely by the criteria of economic rationality and efficiency, for such criteria cannot be treated in isolation. Other objectives will be regarded as equally important, such as persons-oriented development and the freedom of choice, and in some cases may become the principal values. This should turn policy-making towards more humanist ideas of social and economic development.

A few illustrations can be given of similar developments in the theory and practice of social science. One of these refers to the various evaluative techniques called "social indicators," which are meant to measure the extent of social development. Hitherto such techniques have described social conditions in excessively physical and mechanical terms, have disregarded cultural differences and have left the definition of social objectives to professional experts in the field. Perhaps such techniques can be improved to better approximate human needs and aspirations. Another development in this direction is the various attempts seeking to highlight a close relationship between social theory and social practice. An illustration is Brian Fay's concept of critical social science. Scorning the positivist idea of a detached science, Fay argues that we should recognize explicitly the interconnection between social theory and social practice, "such that what is to count as truth is partially determined by the specific ways in which scientific theory is supposed to be related to practical action.[20]

This methodological approach can be strengthened further by adopting some of the ideas developed by Yehezkel Dror in his paradigms for the "policy sciences," which aim at improvement of policy-making.[21] Highlighting the central importance of policies for future development, Dror called for the creation of a science that would provide rational control over policy-making, referring to it as a " societal direction system." He viewed the existing evaluative approaches based on the methods of the behavioral and management sciences as inadequate and narrow: they are too technical, neglect non-economic and non-rational contexts, are too theoretical and therefore inapplicable to policy practice, fail to integrate theorizing of different social disciplines, and have other similar weaknesses. According to him, there is a need for a new policy framework to deal with policy issues, broader in scope than the old one: it would be interdisciplinary, would consider both macro and micro policy levels, utilize the personal experience of policy-makers, examine values and their implications, explore new alternatives and futures, and recognize irrational and extra-rational elements in policy-making (e.g. ideologies, intuition). It would seek to bridge the dichotomy between pure and applied research and move away from mere descriptive and quantitative models to prescriptive and qualitative models.

A similar reorientation of policy approaches is advocated by Bertram Gross,[22] who assailed as " destructive" the decision- making practice in developing countries and called for a change in its perspective. According to him, decision-making strategies in these countries have led to destructive results because they have been largely based on the false premises that " development can be dissociated from the destructively exponential growth"— from how it affects the planet's physical resources and ecological linkages—and because of " the popular myths of entrenched economics" that the GNP is the overall indicator of " progress." These premises, he says, " tend to reinforce the evolving institutions of new-style empire and oligarchy." To overcome these limitations of the developmentalist model, Gross highlighted the need for distributive and redistributive strategies focused on material and nonmaterial growth, larger-scale employment projects and " the fostering and use of more constructive strategies." However, he acknowledged that all such shifts would require " substantial, long-term changes in the sociopolitical structure of developing countries and the world society."

* * * * *

It has been shown that the possibility of overcoming the limitations of technocratic theory and practice and increasing the responsiveness of technocracy to "humanist" ideas and its accountability can be explored by using a variety of strategies. Our discussion highlighted political checks on technocracy through countervailing forces, Galbraith's idea that "the educational and the scientific estate," together with the intellectuals, asserts a type of political and moral leadership that represents "liberal arts" values as against values dominated by industrial or technology-based systems, and Dror's idea of upgrading the educational quality of political leaders to allow them to assume leadership in policy-making and to approach policy from a broad, not merely technical or narrowly "political" perspective. However, these reforms are not deep enough to change the nature of technocracy itself. For this reason, it is proposed that the technocratic concept be reoriented. This aims at breaking the bond between technocracy and the conventional methodology of social science and opening technocracy to more "humanist" and "liberal" ideas of development.

Such a reorientation of "technocracy" is likely to have an important impact on government administrative and decision-making systems. It should generate a more positive interest in social policy and human development and encourage more interaction between technocratic planners and citizens affected by development projects, and should also help to overcome some of the "destructive decision-making" common in developing countries which is due to the adoption of conservative developmentalist models. Moreover, by incorporating humanist goals, the technocratic concept is likely to extend its ideological dimension and appeal. It could thus become more an ideology than a mere technique for handling policy problems and could offer itself as a model for orderly and rational social systems and an alternative to traditional ideological models. In addition, by impelling development thought to turn in a more humanist direction, the reformed " technocracy" should help to overcome the gap between commitment and performance in government policies, which, as the experience in the Philippines and other developing countries as well suggests, has persistently plagued their development efforts.

CHAPTER VIII

Technocracy after Marcos

The advent of the Aquino regime in 1986 marked a series of changes in the Philippine social and political system, but the continuous presence of technocrats in leading government positions and the senior bureaucracy indicates that technocratic influence remained a strong feature in Philippine administration and policy-making. This epilogue compares the extent of technocratic influences in the Aquino regime with that in the Marcos regime.

Even before Corazon Aquino's accession to the presidency, many technocratic planners had expressed their preference for a change in the existing regime. They had become increasingly alarmed at the country's slide into chronic economic crisis towards the end of the Marcos regime. They tended to blame this slide on the contradictions of Marcos's policies as, for example, embodied in the commitment to industrial efficiency alongside the operation of the monopolistic and inefficient crony system. There was also widespread concern over the growing scale of corruption emanating from the center which was a part of this system. Planners like NEDA's Gerardo Sicat, as noted earlier, had already begun to speak out publicly against Marcos's economic policies. Their concern was endorsed in a range of criticisms emerging from the country's international creditors and some conservative political commentators who blamed the continuation of the economic crisis on "the chaotic political and national security situation."[1] They began to support the opposition's demand for a more open, democratic political system, expressing preference for "free and creditable elections," apparently hoping that this would help redress the stagnating social and political conditions.[2]

There were other reasons for friendly relationships between technocrats and the new regime. Technocratic planners were badly needed as advisers to help resolve the ongoing crisis in the economy which had intensified with the murder of Benigno Aquino in 1983 and which had to be brought under control. Corazon Aquino's political inexperience and her ignorance of economic matters made it certain that she would be more reliant than Marcos on the advice of the technocrats. Another reason, which became more apparent later, was Aquino's distrust of the traditional politics of bargaining—her inclination to remain "non-partisan," which seemed to have predisposed her to seek advice from "objective" experts, apart from her loyal personal friends and followers. Thus the accession of Aquino to the presidency was seen as a favorable development by many technocrats. Her economic policies, as thinly elaborated during the snap election, did not threaten the economic orientation they advocated towards the development of an efficient export-industrial sector. In fact, as they perceived it, the Aquino regime presented, in the statement of an ADB report, "an opportunity for taking firm and effective measures to carry through the structural reform process."[3]

The Political Context

Clearly, the role of Philippine technocrats was affected by the new political environment. A number of factors in this environment stand out, the first of which was a major change in the existing political system. A new constitution ratified in 1987 marked a revival of the pre-1972 presidential system, with an elected Congress composed of a House of Representatives and a Senate. This gave a crucial role in public decision-making to a multitude of elected politicians who were also in a position to subject executive policy decisions to public scrutiny and enforce public accountability. In this way also, the actions of technocratic planners could be brought under more direct public control. At the same time, political reforms were accompanied by a reorganization of government administration which was to be introduced throughout the government system, including the Office of the President and the Cabinet. This again could be expected to strengthen public control because it aimed not only at increasing cost effectiveness of government administration but also at enhancing its accountability.[4]

The second factor was a shift in the Aquino regime towards a conservative orientation. The "people power" phenomenon gave the new regime a strong populist appeal. This was partly reflected in the initial composition of the Cabinet, which included three quite progressive members (Aquilino Pimentel, Augusto Sanchez and Ramon Mitra), as well as Joker Arroyo, a human rights lawyer, who became the Presidential Executive Secretary. However, this arrangement did not last long. Increasingly, friction arose between the progressives who were "cause-oriented" in their outlook (which included an interest in civil rights, a "nationalist" orientation expressed in the call to dismantle the American bases, and concern for the poor), and the conservatives who represented big business interests and tended to subscribe to the World Bank/IMF prescription for Philippine development. The progressive ministers and undersecretaries were gradually displaced by more conservative officials. As the popular expectations generated by the 1986 EDSA revolution could not be met, references to "revolution" or "people power" were increasingly dropped from the rhetoric of the Aquino regime. Indeed, this change in policy orientation happened within one year of the advent of the new regime. In the phrase of a contemporary writer: "Aquino is increasingly seen as a leader who is losing the ability to hold the democratic center. She failed to exploit the political potential of the adoring multitudes who swept her to power in February 1986 to push through hazardous but necessary strong policies such as land reform."[5] The effect of this change was to strengthen the technocratic position, for the change occurred in the direction preferred by the technocrats, namely towards an open market economy, privatization of government assets and an export-industrial orientation.

Lack of effective leadership and indecisio making was another characteristic of politic Aquino regime. This is reflected in the frequent c plaints about Aquino's failure to provide adequate government coordination and was exemplified by quarrels within the government, with ministers publicly criticizing one another's policies or refusing to cooperate in major development projects. A well-known example of such a "policy split" was the conflict which arose in late 1989 over a plan to build a huge $40 million industrial project south of Manila, the First Cavite (province) Industrial Estate. Hailed by Mrs. Aquino as "another major step . . . in the country's economic development," the project bogged down because of government infighting and indecision. It became a battleground between government industrial officials and the Department of Agrarian Reform, which wanted to incorporate the land in its land-reform program. This, wrote a journalist then, "is the latest example of how members of the Aquino government—outwardly anxious to attract foreign investment—can't seem to agree on policy and then implement it."[6] A similar case had to do with the issue of the privatization of the Philippine National Oil Co. (PNOC). This was promised by the government and approved in March 1987. According to Vicente Paterno, PNOC's chairman, privatizing Petron was "the key project and could set the tone for all future privatization ventures."[7] Yet this was never to be accomplished. A new chairman, Manuel Estrella, who took over the same month, turned against the proposal, mainly on the ground that since 40 percent of the equity was to be offered to foreign investors the project was against the "national interest." Intense feuding followed between Estrella, who was supported by Joker Arroyo, Aquino's executive secretary, and Jaime Ongpin, then chairman of the government's Committee on Privatization, and his allies, many of whom were Harvard Business School graduates, who rejected Estrella's argument both on nationalist and economic grounds. Despite its promises that it would divest itself of monopolies, the government took no action to change the situation. Petron continued to operate, as one critic put it, "in the heartland of the private sector, out of reach of traditional government strictures and constraints, and self-contained in its flexibility to make and implement decisions . . . [yet] it is at the same time a market participant with privileged and exclusive access to certain accounts and other intangible benefits accruing from state ownership."[8]

Another problem was Mrs. Aquino's inability to provide clear policy direction and adequate backing for her ministers. This led to the resignation of Jaime Ongpin, the Secretary of Finance, a "radical technocrat" and one of her closest advisers in the Cabinet (who committed suicide shortly afterwards). At a different level, Aquino's failure to act manifested itself in the haphazard performance of the Presidential Commission on Good Government (PCGG) established in 1986 to hunt down, seize and return to the people ill-gotten gains of Mr. Marcos and his cronies, as well as in Aquino's much-hailed program to privatize almost all government corporations, including government-owned banks. A writer in the influential *Asian Wall Street Journal*[9] described this situation in 1990 as follows:

> So, many of the free market policies Mrs. Aquino promised back in 1986 lie stranded in the bureaucratic backlog. The big privatization program announced in 1986 has fizzled into an ashheap of excuses for not selling such state assets as the huge Petron oil company, Philippine Airlines and the Manila Hotel—all still run by Mrs. Aquino's appointees.
>
> Price controls warp the markets in basic commodities such as rice, with new controls added as an emergency measure since the latest coup attempt.

The writer called this behavior "wasted energy" and observed that "the tremendous political energy that should have gone into serious reform has been largely frittered away on such populist endeavors as the search for the Marcos loot, the bickering over land reform, the requests for foreign aid, ineffective efforts to stop graft without removing the underlying cause, over-regulation, and now the impending haggling over higher rent for the U.S. bases." He reminded the readers that "Back in the defiant presidential campaign days of early 1986, Mrs. Aquino set out a blueprint for reforms that could indeed lay the groundwork for a secure Philippine democracy. The work toward that crucial legacy has barely begun."[10]

Mrs. Aquino's inability to provide effective leadership was reflected in frequent criticisms of her conduct as policy-maker. She was charged with "avoiding action" in her 1990 "most sweeping" Cabinet change (the change was described as "more of a reshuffle than a re-orientation") and in 1991 with projecting "no sense of urgency" despite serious problems in the political arena and the economy which led to "government paralysis" blocking reforms.[11] A similar criticism was directed at her lack of legislative initiative as well as her inability to mold congressional legislation. Such weaknesses were sometimes used by congressmen as an excuse for their own inactivity. They preferred to direct their criticism at the familiar evils of bureaucratic corruption or misuse of funds but would not take the initiative to introduce politically unpopular changes, and then would blame the government for being inactive. As one analyst put it, "politicians have joined in blaming the government for mismanagement of resources. But no politician is wont to champion the ending of [oil] subsidies,"[12] obviously because such measures are unpopular with the public. The inadequacy of Aquino's leadership was also exemplified in her self-proclaimed "non-partisan" approach to presidential functions, an attempt to disengage herself from the bargaining politics for which she seems to have considerable distaste. Her interpretation of the separation of powers—that laws and policies should be primarily shaped by people's representatives in Congress—has, according to one writer, "isolated the executive branch from the legislature and paralyzed the decision-making process. Instead of presenting a united front, the government has dickered almost over every issue, allowing itself to be seduced by populism and cowered by military and left-wing rebels."[13] Instead of seeking to bridge this gap by means of some formal arrangement between her office and Congress, Aquino vetoed such a proposal. Thus, in spite of her early views on the primacy of Congress in the forming of legislative measures, Aquino was increasingly obliged to circumvent congressional "obstructions" by turning to provincial officials and non-government organizations, with whose support she established the populist Kabisig (Linking Arms) movement, ostensibly to speed up government work.

Other political issues given prominence during the Aquino regime were similar to those prevailing during the Marcos era. One of these was the issue of political patronage. This again became widespread both at provincial and national levels. Political analysts wrote about "the enduring baronial nature of Philippine politics," using the terms "warlordism" and "bossism" for it.[14] The same practices were also widespread at the national level, with the advent of "a new-style cronyism . . . near the top of the Philippine Government,"[15] which did not seem to have excluded members of Aquino's own family (particularly Jose "Peping" Cojuangco, the president's younger brother). On a similar line, Haydee Yorac, chairperson of the Commission on Elections, identified in the early 1990s a general decline in public morality; according to her, "Practices and devices that we used to denounce in the days of the dictatorship have resurfaced with new intensity—except now we do them in the name of democracy."[16] She decried the re-emergence of the "distressing tendency to undermine rather than reinforce the rule of law," adding that "Fraud, terrorism, undue influence and bribery of public offlcals are resorted to without hesitation or remorse to gain unfair advantage."[17] Elsewhere a political analyst sought to highlight the persistence of "the traditional cycle of political patronage." However, he came to the conclusion that this is "far too deep-seated and well-organized and the power in the oligarchy's hands is so great [that] nothing can really shake it."[18]

Another issue given prominence was the performance of the Philippine bureaucracy. This was commonly criticized as being inefficient and "now the biggest in Southeast Asia."[19] According to one analyst, it "is widely viewed as an ineffective implementor of government policy. Often short of funds, low on motivation, reputedly ridden with corruption and short on efficiency, the public service frequently fails to deliver even the most basic services and assistance to the population."[20] It was also assailed for continuing to embody the "nepotism" associated with it during the Marcos era.

Two other factors which played an important role in Philippine politics need to be mentioned: the anti-government activities of radical dissidents, particularly of the New People's Army, and the behavior of the Philippine military. Both had greatly destabilizing effects, but now the military posed the greater danger. Some of its leaders staged a number of coup attempts. They justified their action by pointing to the need for reform in the military and, more broadly, the need to accelerate social and economic progress in the country. In this way they sought to present themselves as advocates of national stability and of the general welfare. It is evident that although in some respects the technocrats' position in the Philippine government system was strengthened, it was also weakened by such factors as indecisive political leadership, threats to political and social stability by the NPA, some sections of the military, and, as under Marcos, by inefficiency and corruption in the bureaucracy and the persistence of the patronage structure in provincial and national politics.

The Economic Context

The technocrats naturally played an important role in the area of economic policy and planning. First, the economic orientation of the Aquino administration after the demise of the Marcos regime reflected both an immediate concern with recovery from the economic crisis as well as concerns with future "sustainable economic growth." The immediate crisis was caused by the inability of the government to meet external debt obligations and had led to the cessation of most forms of foreign credit by the end of the Marcos regime.[21] The structural reforms of the economy envisaged by Marcos but not carried through were now to be implemented. They were aimed at an "industrial renaissance"[22] which was to be effected through reorientation of the economy toward import liberalization, privatization of government corporations, encouragement of export-oriented industrialization, encouragement of foreign investment, restructuring of the financial sector, reforms in the taxation system and other strategies whose objectives were to make the economy more open and competitive, to increase its productivity and international appeal, and to minimize state intervention. Most of these objectives won government endorsement in July 1986 in a 30-page report entitled *Agenda for a People-Powered Development*, which was treated as a blueprint for future national development. This report was in turn based on other reports, prepared chiefly by economics professors from the University of the Philippines, particularly one called *Economic Recovery and Long-run Growth: Agenda for Reform*, which combined proposals for the classical idea of an open market economy and removal of tariffs with ideas of "people-powered" development. The report advocated support for "small producers and the poorest sections of the population through such primary thrusts as improving the incomes of small farmers and a state-funded two-year 'emergency employment program' that would provide immediate jobs to one million Filipinos."[23]

There were, then, basically two uneasily related themes underlying this program for Philippine development. One, reflecting conservative economics, was supported by most technocratic planners and businessmen; the other, reflecting populist-nationalist ideas of desirable development, focused on people's welfare and protection of the "national" interest. The *Agenda for Reform* reflected the two themes, as did Solita Monsod, head of NEDA, a former University of the Philippines economics professor. According to her, the Philippine program for economic recovery comprised "a common commitment to a set of guiding principles, namely, respect for human rights, social justice and poverty alleviation, recognition of the urgent need for growth and efficiency, and a minimum of government intervention."[24] Translated into policy, the "populist" aspect of the program gave prominence in development planning to small and medium rural-based industries and labor-intensive approaches; these were expected to strengthen the mainly agriculture-based domestic market and generate new employment.[25] In practice, however, the technocratic-business approach remained the dominant one. It was observed by Cesar Virata, prime minister under Marcos, that the economy-adjustment program worked out by the Aquino government was essentially the same as that prepared in 1983 with the WB/IMF support; it contained the same "emphasis on agricultural development, the lifting of tariff protection for industry, and the move toward allowing market forces to determine economic efficiency."[26]

Management of the economy was left in the hands of business and technocrats in the senior bureaucracy. Jose "Jobo" Fernandez, respected banker and Central Bank governor under Marcos, was retained in his position. The Harvard-trained Jaime Ongpin, former president of Benguet Corporation, was named minister of finance, while Jose Concepcion, former president of a local food and flour manufacturer, RFM Corporation, and a director of the influential Makati Business Club, was named minister of trade and industry. Solita Monsod, chosen to head NEDA, was one of the few government economists who did not come from the private sector. A similar alliance of business and technocratic leaders still predominated five years later, with Jesus Estanislao, another Harvard-trained economist, heading the ministry of finance (after Vicente Jayme, former head of the Private Development Corporation, an investment bank) and banker Jose Cuisia, holding the post of Central Bank governor. There was, then, considerable continuity during this period in the choice of economic advisers and decision-makers and no major break from the Marcos era.

In this economic orientation private enterprise, not the government, was to be the principal "engine of national growth." In the words of Philippine economist Bernardo Villegas, "Mrs. Aquino wants to activate the spirit of private enterprise that Mr. Marcos nearly killed through crony and bureaucratic capitalism, farm monopolies and tolerance of corruption."[27] The role of the government was essentially seen to be that of providing the infrastructure and other support necessary for accelerating the expansion and productivity of the private sector. The business orientation of this economic thought had several aspects. One was a commitment to the privatization of government corporations. Indeed, these were to be offered both to local and foreign investors at "fire sale prices,"[28] although, as noted earlier, this program fizzled out because of government indecision and bureaucratic interest in preserving the existing arrangements. Another example was the passing of the Omnibus Investment Code in 1987 to improve the investment climate for the private sector. This introduced five-year tax holidays in preferred areas and eight-year tax holidays for pioneer industries, and granted labor-intensive enterprises a tax deduction for additional direct-labor expenses. Among its other provisions were cheaper electricity rates for industrial enterprises and measures to encourage the relocation of industries outside Metro Manila.[29] This trend was reflected in the reorganization of the Board of Investments (BOI), which was to be transformed from a regulatory body into a simplified and more decentralized promotion-oriented agency. Its task was to help the private sector expand business activities and put up new projects.[30]

The increasing influence of the business-oriented element in economic policies inevitably manifested itself in a gradual shift from more "populist" or "cause-oriented" ideas of economic development to more conservative ideas of economic management. One idea popular among the *Agenda* advocates in the early years of the Aquino regime was to stimulate the growth of the economy and people's welfare by Keynesian "pump priming" aimed at encouraging domestic demand and generating employment. In this vein the government announced in July 1986 the setting aside of three billion pesos from the national budget for stimulating the domestic market through the support of small and medium-scale enterprises.[31] But the proposals of the *Agenda* group were sharply criticized by the Philippine Chamber of Commerce and Industry, which viewed them "as 'anti-big business' and simplistic in addressing the country's economic problems."[32] Within a few months this policy thrust was modified, although the pump-priming exercise was not abandoned altogether. (In an unusual move, the IMF had almost completely dropped its usual preconditions for austerity to allow the government to run a higher deficit budget for this purpose).[33] This change was reflected in the government's September economic statement, which affirmed an essentially conservative position: "to keep the economy open, vigorously encourage local and foreign investments, abandon a labor-intensive industrialization program and scrap a commodity-led export strategy [focused on the export of raw materials]."[34] The reference to the dropping of the labor-intensive approach in particular exhibited a more conservative position. It was justified by the argument that "labor-intensive technology is no longer effective especially with the advent of robotics," and that "Labor, in today's international setting, no longer provides the so-called comparative advantage which was true in the '60s and '70s. In the coming decade, the new Government believes capital will be the key to development. The economies that provide the cheapest cost of capital for development will most likely be the ones to succeed."[35]

As in the past, a very important role in Philippine policy-making was played by the World Bank and the IMF, with whom the leading technocrats in the government and senior bureaucracy were closely aligned. The government was vulnerable to the influence of the multilateral agencies because the economy was weak and largely dependent on their assistance for funding its economic recovery program, for meeting its frequent balance-of-trade and budget deficits and riding through other financial crises. It was largely due to the pressure exerted by the IMF and World Bank—consisting mainly of threats not to grant or release further credits—that the government program of "economic adjustment" did not collapse and that the process of dismantling the old system based on protectionism and import substitution proceeded at reasonable speed. Indeed, the story of the progress towards economic adjustment goals was one of constant opposition and delays by the Philippine government, invariably ending in surrender to the WB/IMF "dictates." An illustration of this was the Cabinet decision taken in mid-June 1986 to drop the WB-IMF prescriptions imposed on the country to immediately complete the import liberalization program in favor of its staggered implementation. Yet, less than two weeks later, the government succumbed to the pressure of the two institutions and agreed to remove all quantitative restrictions on imports before the end of the year, although local businessmen "greeted the news with consternation."[36]

As in politics, economic decision-making suffered from a lack of firm government guidance and effective leadership and was marked by indecisiveness, lack of focus and poor policy integration. A criticism frequently raised was that, unlike under Marcos, where decision-making had been clearly concentrated in the hands of the President, "at present the Cabinet members and other heads of agencies are left pretty much on their own."[37] This tended to militate against the possibility of consistency and rationality in policy decisions. Disagreements and clashes on policy issues were frequent and reported in the press. There were, for instance, major frictions between Executive Secretary Arroyo, and Finance Secretary Ongpin. In one case Arroyo criticized Ongpin for imposing new taxes without consulting the people and Ongpin in return attacked Arroyo for blocking his policies. We noted a similar feud between them in connection with the proposal to privatize Petron. More broadly there were public clashes between the advocates of "cause-oriented" ideas of development and business-oriented economists. A well-known instance of it was the disagreement between NEDA's Solita Monsod and two other leading government economists, Fernandez and Jayme, over Manila's debt-management policies. She advocated a "selective repudiation" of debt obligations on the ground that "our aim . . . [should be] a reasonable economic growth first and then [we should] work out a debt-management scheme consistent with that, not the other way around." It was widely interpreted as reflecting her concern that full payment of debt obligations would adversely affect national development and would hurt the lower classes.[38] Fernandez and Jayme, for their part, insisted on honoring all of the country's foreign debts and sought to renegotiate and reschedule them. In the end Monsod resigned because of "honest and irreconcilable differences over economic policies." This move, one policy analyst[39] wrote then, was hailed as "the triumph of a powerful bloc of ministers within government who are closely linked to big business groups . . . Her leaving in effect resolves . . . the tension over economic policy between two camps in the Aquino government [involving] the conflict between the free-market business bloc and those who desire affirmative action to improve the livehood of the poor." He described the victors in this conflict as "the free-market business-oriented bloc, [who] held that economic policies should put priority on business growth and that one of the main obstacles to growth is the bureaucracy's inefficiency and corruption."[40]

At the root of such policy vacillations were the basic differences in economic philosophy described above. According to one writer, they reflect "a policy schizophrenia that has dogged the Aquino administration since its inception. On the one hand is a nationalist-populist standpoint calling for state intervention in the markets—for instance, to raise workers' wages and subsidize fuel prices. On the other hand is the IMF-oriented view, adopted by top economic officials such as Estanislao and Cuisia, that the Philippines' basic problem has been its failure to get prices right."[41] In this conflict Aquino was charged with having "opted for a facile compromise, with disastrous results."[42] For instance, during the oil crisis in late 1990, oil prices were raised to reflect the rise in crude-oil prices, but taxes on petroleum products were reduced to minimize the increase.[43] Another illustration of Aquino's vacillation was her decision to postpone the promised implementation of a package of tariff reforms that was due to start in September 1990. This elicited a strong condemnation from the international business community. One economic analyst, writing in the influential *Far Eastern Economic Review,*[44] declared that this "marks a missed opportunity in the struggle to revitalize an ailing economy," and the editorial article, significantly entitled "Invertebrate economy," viewed this move as a retreat to protectionism from which the regime claimed to be seeking to liberate itself. It complained that "Now that the [tariff-reform] plan is shelved, the Philippines remains stuck in an import-substitution system that has changed little from the 1950s."[45] It added that "That policy worked well in the 1960s, helping to make the Philippines one of the fastest-growing economies in Southeast Asia. But the spurt ran out of steam as manufacturers—multinational as well as local—failed to reinvest profits and remained hooked on imported components."[46] Such economic behavior weakened the rationality of national planning as Aquino sought to accommodate the disparate demands being made upon her regime.

* * * * *

In sum, both the political and economic environment reflected a high level of technocratic influence, although the technocrats' ability to have their ideas adopted was circumscribed by the demands of "practical politics" and was often weakened by Aquino's peculiar management style. This was essentially reactive and lacked a firm direction and a coherent vision. But there were perhaps no major disagreements between her and her technocratic "advisers." When she chose not to follow their advice it was because she seemed to have felt that it was politically necessary to take a different course of action. However, in the end she accepted policies proposed by her leading technocrats and "surrendered," like Marcos before, to WB/IMF prescriptions for reforming the economy. The nature of technocratic influence will be briefly examined in more detail in a number of policy areas—decentralization policy, development policies and planning—that we examined earlier for the presence of technocratic elements during the Marcos regime. The growing interest in science and technology will also be considered.

Decentralization Policy

Under Aquino, there has been considerable focus on government decentralization. The importance of decentralization had also been highlighted during the Marcos regime as part of the process of "democratizing" Filipino politics but, as we noted earlier, the practice suggested the use of centralizing and integrative approaches to increase the capacity of the state to manipulate "decentralized" units; the technocratic element in this policy was strengthened through national planning and control. During the Aquino era a commitment to decentralization was undoubtedly more genuinely held; it expressed itself in many respects. It is reflected, for instance, in the foreword to the *Philippine Development Plan 1987–1992*, which, according to NEDA's chief planner, Solita Monsod, "includes a regional development framework which is based on the Regional Development Plans and Investment Plans approved by the Regional Development Councils"; the entire Chapter 2 was devoted to a "Regional Development and Physical Planning Framework." Increased regionalization and local autonomy were to be achieved through "devolution of powers, programs, resources, and capabilities"(p. 402); they included the delegation of more substantive and administrative authority to regional and local governments, increases in local budgets and decentralization of decision-making, as well as enhancing the capacity for local planning. In this connection the powers and functions of the Regional Development Councils were to be expanded to enable them "to take a more influential and substantive role in regional policy and decision-making."(pp.204–205)

Nonetheless, alongside these commitments to regional and local development policy, the centralist outlook has been retained in many areas. At the outset the plan made it clear that the regional perspective was to be related to the interest of the whole nation. "The fundamental goal of regional development," it stated, "is the minimization of disparities, without prejudice to the optimum realization of the regions' growth potentials . . . It is expected that the present strategy . . . will rectify the imbalance among and within regions."(p. 49) To achieve decentralization it was proposed that government ministries be encouraged to strengthen their offices in the regions, that clear guidelines and standards be formulated for local governments, and that the mechanism for interagency coordination of national policies and programs affecting local governments be improved. Furthermore, improvements in national supervision and assistance to local government units were given emphasis.(p. 405) This suggested a strong central focus in planning, especially as the capabilities for local planning were, it seems, conceived as supplementary to traditional NEDA regional planning. Thus, the bond between local and central systems remained close as in the past, with the goals of local development being perceived in terms of technocratic ideas of comprehensive national development and planning.

The same orientation is reflected in a 1989 review by NEDA's Solita Monsod, describing the " Decentralization Thrust" in national planning.[47] She referred to " the move toward meaningful and functional decentralization" and identified a number of strategies developed by NEDA in this direction. They included the reorganization and strengthening of the local development councils, improvement and updating of the regional planning framework, improvement of the capabilities for regional planning and processing of projects, periodic reviews of regional budgeting, integrated area development and regional development investment programs.[48] According to her, NEDA also increased its linkages with implementing agencies through better mutual consultation and through coordination in monitoring development plans; adopted a regional monitoring and evaluation system for its regional offices; and introduced evaluation of " sectoral, regional and project level performance . . . to determine whether the desired level of public investments is consistent with macro-economic targets."[49] A conspicuous feature of economic planning under the Aquino government has been the Manila-centric focus of much economic activity. As the government has struggled to maintain economic services in Manila, infrastructure in other parts of the country, notably Mindanao, has decayed. One factor in the government's push for greater provincial autonomy undoubtedly lies in a desire to emancipate the national capital region from the burden of providing services outside it.

Some aspects of the decentralization policy developed under Marcos were then carried over into the Aquino administration. Nonetheless, major changes in this policy could be initiated only by Congress. A variety of proposals were made by congressmen on this issue, including calls for the creation of a federal system. Five different versions of a new Local Government Code were submitted to Congress for the period up to 1990 and it was at one point anticipated that it would take several years before a decision on them would be made.[50] This delay was explained mainly by the inability of Philippine congressmen to decide on the form which such decentralization should take and their reluctance to lose control over funds. The vacillating performance of Congress thus cast doubts on whether the existing system of centralized management, based on technocratic ideas of comprehensive approaches to planning and development, could be changed. The eventual passage of a local government code which has significantly enhanced provincial and municipal fiscal autonomy will doubtless re-awaken the interest of many national political actors in a regional political career.

Development Policies

Earlier we discussed the impact of Marcos's development policies on forest and watershed resources, and noted some of the problems created by the commitment to technocracy in this area. Over ten years later the Philippine forests were still in a state of crisis. One government official writing in 1989 drew attention to the shrinking area of old-growth forests (now only 900,000 hectares out of 5.5 million that are " technically classifiable as forest areas") and warned that without immediate action commercial-grade timber would be lost in ten years.[51] He likened the efforts to preserve forests to " fighting to preserve our heritage . . . in the rural and uplands areas, where our ancient and once monumental forests are being destroyed and depleted by the relentless onslaught of commerce and human encroachment."[52] The same feeling of crisis was projected in the development plan for the period 1987–1992. This referred to deforestation and soil erosion as being "the largest and most serious environmental problems in the country."[53] This situation was blamed largely on the uncontrolled exploitation of forests, particularly commercial logging.

The government recognized that urgent action was needed to ensure forest protection and conservation. The problem was to be treated in a technical way, as one of effective use of resources. A number of strategies were proposed in this connection, which aimed at minimizing the rate of forest destruction, reducing pressure for upland migration, preventing illegal logging, encouraging reforestation, and developing and proper utilization of timber resources.[54] There were programs aiming at strengthening forest law enforcement, continuation of the log export ban, reforestation and rehabilitation of critical watersheds, as well as long-term programs for conservation, preservation and administration of national parks and games-and-wildlife refuges. An original proposal was the creation of planned forest communities which were to consider both "the social interest of man, and the ecological imperatives of nature."[55] In addition, the role of "a free and vigilant press" was said to be important. According to President Aquino, this "could monitor the granting of logging licenses and concessions so that the interests of small settlers are protected and the monopolies in the exploitation of natural resources are avoided."[56]

In this technical context of forest management the human aspect of the policy was not given sufficient recognition. The issue of land ownership, of forest occupants, particularly the cultural minorities, and of ancestral land was virtually ignored. As noted earlier, the entitlement to ancestral lands had always had a constitutional basis but in practice this principle was not applied and the "Regalian doctrine" (maintaining that all lands uncovered by official titles are presumed to be owned by the state) continued to be the basis of government policy.[57] For most mountain and forest dwellers, such titles were not accessible because of the complexity of legal and bureaucratic procedures involved in securing the necessary documentation.[58] The situation in this respect became increasingly difficult under martial law and the law did not change under Aquino. Lands policies continued to be based on the " Regalian doctrine" and there were no new legislative measures giving explicit recognition of land ownership to all cultural minorities or long-term occupants.[59] There were a number of proposals to rectify this situation but these came from outside the forestry bureaucracy. The defenders of customary rights advocated giving explicit recognition to ownership of land to all cultural minorities and long-term occupants and allowing lands to be administered by NGOs concerned with forest development. These proposals were based on the belief that this would not only reflect the actual legal position of the cultural minorities but would also be a most effective means for protecting land against further erosion and destruction as the owners would have a personal interest in maintaining the land in good condition.[60]

The issue of forest conservation and development was dominated by a government bureaucracy, namely the DENR (Department of Environment and Natural Resources), and was treated in a rather mechanical, technical way. The justification given by this agency in 1987 for this approach was that the land erosion, siltation and loss of soil fertility could be avoided only " through government control and regulation, a fact not possible when the land becomes private property."[61] This position, however, ignored the potential contribution to forest conservation and development by permanent residents and overlooked the inability of the department to deal with major problems because of " inherent manpower and financial limitations."[62] Moreover, the attitude of the bureaucracy can perhaps be explained by its vested interest in perpetuating the existing system: it stood to lose control over some 25–50 percent of the land currently under its jurisdiction if recognition of customary ownership was given legal force. In addition, the difficulty of introducing logging reforms suggested the possibility of close cooperation between forestry officials and the big forest concessionaires. An example of this was the inability of Fulgencio Factoran, DENR chief, to enforce anti-logging policies within his department because his directives were undermined at the lower bureaucratic level. As one contemporary report put it, "Directives get ignored at field level where private interests often take priority over formal chains of command."[63]

The DENR's technical orientation was reflected in many government documents, including the development plan for the period. These indicated that former concerns for forest inhabitants, cultural minorities and ancestral lands were disregarded. These concerns were not even mentioned in a 1989 statement of government policy by a high DENR official,[64] yet they were crucial to those affected by this policy. Similarly, they were not addressed in DP 1987–1992, where only two marginal references to them appeared in the context of dealing with efficient forest management and planning:

> Efforts to distribute lands more equitably, however, were diluted to some extent by exceptional releases of vast estates to private and multinational corporations which resulted in the dislocation of the original settlers, including cultural minorities. Moreover, the absence of clear-cut policies and guidelines on land use and land tenure issues involving the existence of idle lands, ancestral lands, access of foreign investors to public lands and the ecological-land balance ratio resulted in the inefficient and suboptimal utilization of land resources and contributed to the social unrest in the countryside.[65]

It is evident that the omission of such issues was not accidental. It reflected value commitments in a technocratic direction. Human concerns continued to be replaced by technical concerns. Viewed from this perspective, the refusal to recognize traditional land ownership appears to have been the aim of the planners, who were eager to retain control over forest resources and to maximize their utilization. It is interesting that the only place in DP 1987–1992 where local people were mentioned was in the concept of " integrated social programs," which was defined as involving " the provision of security of tenure over forest lands actually occupied by upland farmers and assistance in the development of said lands through agroforestation."(p. 134) This concept was elaborated in the 1989 policy statement, where it was related to ecological concerns and referred to as " forest cities" or " planned forest communities."[66] It was stated that

> the centerpiece of such a stategy is the maintenance of a healthy forest area, planned and programmed for growth, harvest, and sustainability. Such a forest should be planted, maintained and nurtured . . . [to maximize] the utilization of available forest space . . . such planned and carefully nurtured forests spread throughout the country could form the basis of upland industrialization, with the forests and its products as a common denominator. Upland dwellers can be settled around such areas, in well-delineated residential plots; . . . such projects should be developed through an integrated area approach . . .

It is evident that the primary concern underlying the DENR's social communities scheme was better management of forest resources and that people involved in these projects were considered in that technical context. Again the question concerning their rights to ancestral land or occupancy was not mentioned in these documents.

The issue of ancestral domain, however, surfaced repeatedly in the discussion surrounding the attempt to set up an autonomous region in the Cordillera mountain provinces. The call for autonomy had initially come from Left-aligned cause-oriented groups towards the end of the Marcos period. It was taken up, in the early flush of reform, by the 1986 Constitutional Commission, which adopted a provision requiring Congress to frame an organic act for this purpose. But with the restoration of traditional political practices under the Aquino regime, the Left lost control of this initiative and the measure was eventually defeated in a regional plebiscite. The defeat was strongest in the populous Ilocano areas of the Cordillera, where the ancestral domain provisions excited nervousness among non-Igorot land holders. There was also some well-financed opposition from lowland-dominated business and mining interests which feared loss of mineral rights. The DENR itself behaved predictably by framing its own forest management proposals aimed at undercutting any organic act provisions which seemed likely to yield control of forest resources to the indigenous occupants. Other pressures were evident during the congressional drafting phase, with individual senators following their own political agenda in pushing for provisions requiring an absolute logging ban. The result was an Organic Act which accommodated so many non-Cordilleran concerns that it was finally defeated even in the majority of Igorot municipalities.

Development Planning

As we have seen, development planning was an important indicator of emergent technocracy during the Marcos regime, where the process of planning became increasingly more comprehensive and more technical and reflected preference for accelerated development and economic and administrative rationality. These themes were continued in various aspects of DP 1987–1992, which was to "serve as the framework of development policy-making in the next six years" (Foreword).

In its overall format, the plan changed little from the preceding document (DP 1978–1982, including DP 1978–1987). Indeed, both plans were divided into the same six parts and had the same number of pages (443–444). The sole innovation was in Part Four, which was expanded from "Infrastructure Support" to "Infrastructure and Technology Support" to give emphasis to the role of science and technology in national development. Again, the economic and infrastructure sections predominated; they were followed by the social and financing sections; the concluding section focused on plan implementation and development administration. This indicates a continuity in the approach to national planning.

The principal goals, described in the early part of the plan, were defined as " economic recovery in the short run and sustainable growth in the long run"(Foreword), more explicitly in terms of four principal goals: alleviation of poverty, generation of more productive employment, promotion of equity and social justice, and the attainment of sustainable economic growth.(p. 11) Special emphasis was given to the need to institute structural reforms within the economy and to the need " to fight poverty." This, the plan stated, " proceeds from the commitment to the principle of upholding the human right of every person to a decent life." The plan promised to make the poor more " self-reliant," to dismantle past privileges to " selected groups" and encourage " fair competition":

> This calls for the transformation of the poor into self-reliant and productive citizens of actively participating in the total development effort. The government is determined to move away from granting privilege to selected groups to a system which emphasizes equity, efficiency, and social justice. Cronyism shall be eliminated, and monopolies shall be fully dismantled and replaced by greater reliance on free and fair competition.(p. 3)

In addition, the rhetoric of " people power" was invoked; sectoral policies were said to be based on the overall policy framework embodied in the " Policy Agenda for People-Powered Development" and sectoral targets to be set " considering our people's needs" (Foreword). The urgency of bringing about more rapid development was highlighted because of the economic crisis which started in 1983, increasing poverty levels, rapidly growing population (at 2.5 percent per year) and the need to restructure the economy because of its continued inefficiencies (pp.3–11).

As in the previous plans, prominence was given to economic objectives. According to the plan, " The resurgence of economic growth is the most immediate task" and its principal tool will be " a demand-led, employment-oriented and rural-based strategy."(p. 3) The plan estimate

ture economic growth did not indicate a major departure from the past economic trends. Figures for the structural distribution of (gross) domestic product for 1986–92 indicated a decline in the agriculture sector (from the 1986 estimate of 29.5 percent to 27.5 percent annual average for the period), a growth of the industrial sector (from 31.3 percent to 33.2 percent), including manufacturing (from 23.9 percent to 24.2 percent), and the services sector remaining unchanged at 39.3 percent (p. 18; Table 1.4). "Social development" targets focused on "poverty alleviation, the promotion of social justice and increased productivity"(p. 28) and included health, nutrition and education. Targets were also defined for poverty reduction. At the national level this was expected to fall from 59 percent in 1985 to 45.4 percent in 1992 and to be focused on the provision of basic human needs "to ensure that these do not fall below maximum requirements" (p. 32, and Table 1.9 on p. 33).

Strategies and policies used for planning were again conceived technically. They aimed at generating employment, optimizing the use of resources, stimulating economic recovery and production, expanding export by making use of the comparative advantage in labor and agriculture, and so on. As under Marcos, the role of the private sector was given prominence; it was to serve as "the prime mover of development" (p. 39). What was new was an emphasis on the role of science and technology in development. Highest priority was given to agriculture and rural development, "in keeping with the goals of alleviating poverty and increasing employment opportunities and incomes in the rural sector" (p. 42), and in this connection it was indicated that an employment-oriented, rural-based development would be implemented in the medium term (p. 32). An important aspect of the development strategy was close international cooperation. The plan stated: "Economic diplomacy shall be directed toward expanding market access to a wider variety of products and services. This shall be implemented by a . . . comprehensive export promotion strategy" (p. 48).

The pervasive technical orientation of the plans is illustrated in the section on industrial development and on the plan implementation. Industry was expected to grow rapidly at an annual average rate of 8.8 percent and to generate an increase in employment at annual average rate of 8.3 percent (p. 45). A package of policies and strategies was promised, " anchored on the principles of comparative advantage and industrial efficiency." (p. 146) These included sectoral programs to revitalize and rationalize the industry sector; promotion of micro cottage and medium enterprises; promotion of investment; trade liberalization; promotion of domestic trade; regional dispersal of industries; promotion of exports (" to effect a more open trade regime"); and international cooperation (" to harness and maximize the current and potential benefits from existing liberal and multilateral relations") (pp. 151–157). These policies were to be supported by appropriate programs and projects. Some of these were ongoing programs which focused on industrial promotion and included a number of lending projects, the Agro-industrial Technology Transfer Program, Export Industry Modernization Program, and promotion of investment, of domestic trade and export. There were also new programs for small and medium industries, private enterprise development, corporate strategy formulation, industrial relations development, and so on.

In this management-oriented planning context, the need for additional social development was recognized in the targets mentioned earlier, but the issue was treated in a technical and conservative way. For instance, family welfare was to be enhanced by the promotion of small families or reduced population growth rate and " responsible parenthood" (p. 44), and manpower development programs were now to be focused on entrepreneurship and skills training (p. 45). Much calculation was done to show how poverty was to be reduced, and the associated concept of basic needs was given prominence. It was promised that " relative poverty will be minimized through the government's development strategy particularly focused on the poverty groups."(p. 32) This seems to have referred to the development strategy used for the whole country, which, for the short term, aimed at generating employment particularly in the rural areas "through the construction of small-scale, labor-intensive infrastructure projects such as feeder roads, communal irrigation systems, school buildings, and rural water supplies,"(p. 35) and through other labor-intensive development projects.

The importance of plan implementation and development administration was highlighted in the concluding section. It was contended that past attempts at instituting administrative reforms failed "to produce a significant impact in terms of enhancing the efficiency, effectiveness and responsiveness of the bureaucracy," and this was explained largely by "the lack of sustained political will to achieve the true intent of such reforms." But such reforms becáme more urgent because of "the wide-ranging changes in our political, economic and social life bought about by the February revolution [which] exert stronger demands . . . on the new government."(p. 39) Among proposed major reforms were: improving the government's responsiveness to the public, rationalizing the government corporate sector, decentralization (which included devolution of power, programs, resources and capabilities, and complementary local, central and regional reforms), increased professionalization of the civil service, development communication (including a free flow of information between the government and the general public) as well as improving institutional processes for plan implementation and monitoring. These policies were accompanied by appropriate programs and projects. (pp. 396–418)

In sum, development planning during the Aquino regime followed very closely the pattern established by technocratic planners during the Marcos era; it was focused on the management of national resources. Prominence was given to the centralization of resources, comprehensive and integrative approaches, and the rationalization of systems. Some references were made to the "rights" of persons and "people-powered development," but they were largely used in a rhetorical way. Policy proposals made for reducing poverty and increasing employment by rural-oriented programs were defined in essentially economic planning terms. The references to cultural minorities or ancestral lands which had been present in the earlier plans were now dropped, again reflecting the continuing technocratic approach in dealing with social issues.

Science and Technology

A preoccupation with science and technology in increasing the productivity of systems continued to be a major aspect of development planning during the Aquino regime. Science and technology (S & T) objectives, strategies and programs were defined in many documents, particularly in Chapter 11 of the *Philippine Development Plan 1987–1992*, entitled "Science and Technology," and in a World Bank report published in 1987, dealing with "Issues and Policies in the Industrial Sector."[67]

The documents criticized the inadequacy of institutional arrangements in the past, despite the fact that "the country has considerable potential for advancing average technology based on its well-educated population."[68] It was contended that the technological element was treated merely as "peripheral to strategic [national] decisions."[69] Another criticism of the past practice was that institutional support for technological modernization was weak and was based on trade and industrial policies "which distorted the choice of technology and provided little incentive to reduce costs."[70] They were largely intended to lend support to protected domestic industries and aimed at product differentiation rather than cost reduction.

The new strategy proposed to treat technology as an integral part of national development planning. Its main objective was to utilize imported technology to meet the need for rural development and economic growth and "to build up an indigenous capacity for science and technology towards technological self-reliance."[71] Government support concentrated on technological infrastructure, which included the provision of technological information services, technical education, establishment of Research and Development (R & D) centers, and the development of standards and patent protection.[72] The government was also expected to provide a favorable policy environment that " encourages the right kind of technological effort," to encourage more directly technological innovations in the industrial sector, promote increased participation of firms in R & D, develop industrial consultancies, and disseminate existing technology, particularly to small and medium-scale firms.[73] More broadly, the government support of technological modernization aimed at upgrading both products and processes (focused more on cost reduction) and at developing or utilizing industrial technology " for successful export as well as to meet the sharpening competition from imports."[74]

The government's faith in the efficiency of science and technology was comprehensive. S & T were to be utilized to upgrade the modernization process in rural and industrial sectors and to generate new employment. Both the public and private sectors were expected to participate, the former providing mainly an infrastructural support. A package of policies and strategies was proposed in this connection: generation and diffusion of employment-oriented technology, particularly in agriculture and industry; development of local materials and indigenous technologies; employment of R & D directed toward commercialization; adoption of imported technology (rather than the creation of new technology); dissemination of appropriate technologies; increasing accessibility to S & T information and services; development of high quality S & T manpower in growth areas; and expansion of education and training in this field.[75] These policies were then supported by appropriate programs and projects.

The preoccupation with science and technology during the Aquino regime strengthened the direction of more technical approaches to national development in Philippine development planning. In the context of planning, the emphasis on the application of S & T methods and techniques reflects the continuing attempt by planners to increase the economic rationality of the government system by using concepts that are intimately associated with technocratic values and practices.

Continuity and Changes

Our survey of the Aquino era indicates that the technocratic mode of thought and action continued to be a strong factor in Philippine administration and policy-making. The presence of technocratic influence was particularly extensive in Aquino's economic policy, where technocratic plannners, together with business interests closely allied to them, have played a dominant role in the management of the economy. They have been strongly represented in the government and the senior bureaucracy, where their professional advice was needed to deal with the economic crisis during the immediate post-Marcos phase and to plan for sustained economic growth. In more detail, the presence of approaches indicating a technical-technological orientation can be identified in government decentralization, the policy of forest conservation and development, Philippine development planning, and in the continuing preoccupation with science and technology.

In sum, government decentralization was not necessarily conceived as leading to increased autonomy for local units but rather to the subordination of local concerns to national development priorities, and, in particular, to the easing of the provincial development burden on the national capital region. The authority of line agencies at local levels was to be strengthened, a move conceived essentially in the context of national planning and aimed at increasing the overall efficiency of the national system. The policy of forest conservation and development was conceived in terms of the optimum management of natural resources, and issues concerning forest inhabitants—the right to occupancy, ownership of land, particularly by the cultural minorities, the right to ancestral lands—continued to be disregarded. Forestry bureaucrats insisted on complete central government control over the use and disposal of forest lands, including logging. A similar technical approach was projected in Philippine development planning. The plan for Aquino's period showed that neither the format nor the content changed; aggregate industrial growth still remained the principal strategy of growth. As in the past, rationalization of systems, comprehensive planning and integrative approaches were given prominence. The management element was given emphasis throughout the document. It included proposals for optimum utilization of resources and, in the section on development administration, proposals for the development of adequate planning and administrative capabilities through appropriate training. References were made to "people-powered" development and "social justice" to indicate a moral commitment, but, as in Marcos's plans, such ideas were translated into the language of economic and technical planning which tended to rob them of meaning.

The preoccupation with science and technology was more explicitly stated than it had been under Marcos. It focused mainly on the application of science and technology to production. The role of the state would be to disseminate information and provide training in this area and strengthen research and development directed towards commercialization. The discussion indicates a continuity in technocratic influence and practice from the Marcos era to the Aquino era. At the same time under Aquino the sphere of technocratic interest has expanded in new directions.

The position of Philippine technocrats was inevitably affected by the more open political environment following the demise of the Marcos regime. Under Aquino, they were exposed to criticism by politicians and had to be sensitive to them, and were not protected by the absence of countervailing forces. Displeasure of congressmen or of influential interests could lead to pressure for their displacement from government positions. This was demonstrated by the sacking of Secretary Jaime Ongpin, who was removed from office because of political pressure, not incompetence. On the other hand, their position was in some way stronger than before. They were badly needed as experts capable of putting an ailing economy on the path of recovery and growth, and they obtained strong backing from the World Bank and the IMF, which were in a position to greatly influence the course of Philippine development by their power to grant or withhold financial assistance. Part of their strength was derived from the fact that their goals were relatively clear and specific, unlike the goals of their political mentors—a free market economy, dismantling of tariffs, privatization of government business, and so on.

The position of the technocrats was in many ways enhanced by this new environment. Theorists of technocracy often claim that technocracy flourishes best in authoritarian or military systems that profess a conservative developmentalist outlook and that there is an ideological convergence between technocracy and such systems. This is not merely because both profess commitment to "development" but because they believe in strong executive authority and are disdainful of liberal politics.[76] The Philippine situation does not lend support to this thesis. Authoritarian systems may give additional strength to efforts at rapid development but they may also act as a straightjacket on further development. This was evident in the Philippines during the authoritarian Marcos regime. Rule by presidential decree during this period led to the increasingly capricious use of centralized power, alongside the formation of new inefficient monopolies in a new system of political patronage, including "crony capitalism." For this reason, leading technocrats like Sicat sought to distance themselves from Marcos's system.

But while there are advantages to a system which is politically and economically more open because it allows more "flexibility" for exploring alternatives and so gives better opportunities for improvement, the effectiveness of such "open" systems depends, to a large extent, on the ability of their leaders to provide firm political guidance and direction. In the Philippines this has been lacking after the change in government because of Aquino's weak management style, which caused a great deal of frustration and irritation among the country's technocrats and other leaders. The effectiveness of more open systems also depends on the conduct of national politicians. Their somewhat subdued criticism of technocratic planning seems to indicate that after the EDSA revolution most of them were more prepared to accept ideas propagated by the business-technocratic bloc than their counterparts in the rural aristocracy-dominated Congress of the pre-martial law era.

The first Aquino government was composed of different ideological hues, including human rights activists, nationalists, conservative politicians and businessmen, technocrats and others; it was an artificial creation brought together in the aura of the EDSA revolution and the demise of Marcos, but was unlikely to be sustained once the revolutionary ethos lost its power to unify people and more pragmatic interests began to reassert themselves. The "Left" was represented by a dwindling quota of cause-oriented activists who openly supported Mrs. Aquino during the snap election and wanted changes in the social structure of the society, while the conservatives were mainly concerned with repairing the damage caused by Marcos to the old liberal system. Aquino's own outlook was conservative: her idea of the EDSA revolt did not seem to go much further than the perception that what was needed was a restoration of a liberal-democratic system, with social change to be left for the elected people's representatives to determine. As these representatives were mostly conservative in outlook and included many politicians associated with Marcos's regime, little structural change could be expected from them. Thus there was an inexorable shift in government from the "center" to the "right" of the political spectrum, represented mainly by the traditional rural interests and the big business interests, and given impetus by the continuous threat of a military coup.

In this political and economic context some conflicts and tensions inevitably surfaced among the technocrats. The old debate of the Marcos era between populist-nationalist commitments and transnationalism was revived in a new form. The populist theme, advocated by NEDA's chief Solita Monsod and supported by U.P. academics, propagated ideas of broad social and economic progress. While the free market was also one of the main objectives, development was to be "balanced" by more people-directed concerns. In planning, this included the use of strategies that sought to alleviate poverty, generate employment, encourage small and medium-size businesses and industries, particularly in rural areas, give more protection to "national interest" in industrial development, and revitalize the rural economy as a start of more robust national industrial expansion. These proposals echoed Aquino's original commitments to "people-powered" development. The second theme, advocated mainly by leading technocrats (except Monsod), "Makati" big business interests and traditional exporting groups, reflected a conservative developmentalist outlook, which was associated with ideas of "structural adjustment" along WB/IMF lines, export-oriented industrialization, removal of obstructions to free trade, the dismantling of tariffs, and so on. Monsod's resignation from the government, as noted earlier, was widely viewed as signalling the victory of the free-market business-oriented bloc. During the Aquino era, Philippine technocrats were not monolithic, and Aquino's preference was ultimately for a development approach that was favored by Philippine big business and technocrats. As in the past, the leading technocrats were convinced that such an approach provided the surest way to revitalize the economy and to bring about sustainable economic growth. It may be noted that the support given to the developmentalist conservative position was not always clearcut. There was an important division between rural interests, well represented in Congress, and the big business sector. For example, the World Bank promoted quite a radical reform but was blocked in Congress. Much of the "rationalization" demanded by the Makati business elite and the "nationalist" bloc were counter to the interests of the provincial elites—an old pre-martial law conflict, in which Marcos backed the Makati elite, but in a haphazard way.

The influence of technocracy during the Aquino era has had a considerable impact on the reconstruction of the post-EDSA Philippines. This has been due to the central position of Philippine technocrats in national planning and to their pursuit of certain ideas that were coming to dominate the Philippine economic scene: that of the free market, export-oriented industrial development and the like, which were also supported by the WB/IMF ideas of "economic adjustment" and growth. We noted earlier that with the advent of the Aquino regime the technocrats felt that they had a chance to really do what Marcos had promised but never delivered, namely to introduce a more rational social and economic system. Under Aquino they seem to have had more opportunity to put their ideas into practice. The impact of policies based on these perceptions cannot be underestimated. They contributed to various structural changes in the economy from the late 1980s onwards, which in turn resulted in considerable structural changes in the social forces in the Philippine society. It has been observed, for example, that the country had moved from its former dependence predominantly on raw material exports to exporting manufactured products; while in 1976 these accounted for 22.1 percent of total exports, in 1986 they represented 61.3 percent of exports.[77] A major change was also experienced in the rural areas. By the year 1988 there were about two million tenants, but the labor-force totalled 6.5 million.[78] According to Jeremias Montemayor, a peasant leader, "wages and conditions are now bigger issues than tenancy and land."[79] This situation considerably affected the class composition of the rural sector and eclipsed in importance Aquino's agrarian program. A writer in *The South* made essentially the same point in 1988: that Aquino's agrarian program was relatively irrelevant to the solution of social conflicts in the Philippines' rural areas. The program, he said, was "meant to satisfy the aspirations of an emerging middle class and agro-industrial elite," for "land is no longer the key to wealth and power. The chief source of patronage is the government, and commercial figures with access to government contracts and favors have replaced the landed gentry at the top of the pile."[80] Thus future social conflicts were likely to occur between old landlords and these emerging classes rather than to include the peasants. The writer identified these classes as comprising "rice merchants, truck owners, small industrialists and suppliers of fertilizers . . .

along with professionals, local officials and those landlords who have diversified into business." And he adds, "As landlords have died, their holdings have been divided among their children—a process which has further tipped the balance in favor of the emerging classes."[81]

A similar observation was made by a social analyst writing in 1990. He diagnosed radical changes in the Philippine economy which he saw moving in the direction of "plutocratic" big-business groups or conglomerates. He also rejected as a myth the widely-held belief in American dominance of the economy, suggesting that this had missed more recent developments:

> While land ownership remains a simmering social problem . . . the focus of economic development will not be on changes in land ownership. It will be basically on how the big-business elite, in a situation of little government interference, foregoes the Philippine elite's traditional proclivity towards conspicuous consumption, and instead use profits both for their employees' welfare and to accumulate capital for industrial development.
>
> Nationalist rhetoric, particularly that which claims that "US monopoly capital" completely dominates the country's economy, has missed the vast change that has occurred in the Philippines since the 1950s. There is a levelling off, and even decline, of colonial-type US companies' dominance. Much of the Philippine economy's growth in the past two decades has been made by local big-business groups, some of the biggest of which cleverly linked up first with new US firms, then with Japanese, and now with Taiwan and Hongkong capital.[82]

This discussion draws attention to the considerable changes in the social and economic structures, which, while latent in the ostensible export-oriented policies of the Marcos regime, were really only confirmed in the Aquino era. This confirmation was partly a consequence of technocracy-advocated policies through national planning. Philippine technocrats helped to shape Philippine development in a new direction, which included the formation of new social groups or interests. This did not, of course, necessarily mean that such changes led to a better ordered society, for, as in the past, the new benefits were distributed unequally and so were likely to create new social problems and instability.

Controlling and Reorienting Technocracy

Developments in the Philippine government during both the Marcos and Aquino eras indicate a high level of technocratic influence in policy-making. The emergence of this influence has been traced to attempts to "accelerate development" in accordance with developmentalist models of modernization. The Philippine experience confirms the original suggestion that the emphasis in the new states on "dynamism in development" marks an important stage in the emergence of the authority of technocratic thought and has led development thought to turn in a technocratic direction. It has been shown that technocracy has had considerable impact on the structure of the economy through technocratic ideas initiated by national planners and that it contributed to changes in social structures. For example, in the early 1990s the issue of land reform and of agricultural tenancy appears to have diminished in importance in debates on Philippine development, while the issue of agricultural workers has acquired new importance. In addition, new social and occupational groups have gained prominence. Some of these changes had been latent in the Marcos era but were confirmed only during the Aquino era.

It has been shown that technocracy can have adverse effects on social systems. With their overemphasis on technical values, technocratic planners tend to treat other values as being of secondary importance and, for the sake of greater rationality and efficiency, adopt methods of social engineering to control and "manage" people in accordance with technocratic theories of rational progress. As we have discussed earlier, the technocratic idea of managed societies can be accompanied by the use of repression and threats to human freedom, particularly for liberal systems where the freedom of choice and expression are viewed as necessary conditions for desirable existence.

Technocracy can also have an adverse impact on human and "humanist" goals of development. In the Philippines these goals were given a high profile in Marcos's ideology of the New Society through concerns with "basic human needs" and the "City of Man." Likewise, Aquino's use of "people-powered" development clearly signalled a commitment to humanist values based on real participation and concern for the popular welfare. Yet Aquino's "political revolution" failed to turn into a "social revolution," and in time her concern for development assumed the form of concern for productive output, as it did under Marcos. This was due at least in part to the developmentalist model preferred by technocratic planners, which professed "growth without equity." Thus technocracy failed as an agent of people-focused planning and development. This weakness of technocracy was reflected in the inability of national planners to deal with the issue of poverty. In a confidential 1988 WB report, entitled "The Philippine Poor: What is to be Done?" it was stated that "there are more poor in the Philippines today than at any other time in recent history"[83]; the incidence of poverty was blamed on "unequal asset ownership, rapid population growth and lack of new jobs" and the government was criticized for having "failed to address these structural issues in the past and, if anything, has contributed to income concentration."[84] But as a believer in comprehensive planning, the World Bank should have perhaps blamed technocratic planners for their inability or unwillingness to deal with this issue. Another illustration can be found in the frequent clashes in the early years of the Aquino regime between "cause-oriented" advocates of development and conservative developmentalists propagating changes through export-oriented industrialization and other similar economic means. These two approaches could not easily be reconciled by Philippine planners operating on technocratic assumptions.

The Philippine experience also draws attention to the adaptability of technocracy to different political systems. It is frequently believed that technocratic planners show a preference for authoritarian systems, yet in the Philippines, technocratic "goals" flourished under Aquino's more liberal regime and atrophied under Marcos's authoritarian regime. There is, then, no necessary link between technocracy and authoritarianism. We have noted the ostensible anti-political tendency of the technocratic position, yet even in the 1960s Philippine technocrats were not openly hostile to the prevailing liberal system. They saw the advantage of political liberalism for economic liberalism, which they believed was the most productive of systems, and only wished to make it more "rational" by an injection of technocratic planning. Their hostility was directed at the existing undisciplined practice leading to social and political instability and insecurity which made rational and dynamic planning impossible. Hence, they initially welcomed the martial law regime because it seemed to promise stability and social progress; it also showed itself more responsive to ideas of technocratic rationality and planning. Later, disappointed with Marcos, they welcomed Aquino's regime for essentially identical reasons. Their expectations seem to have been more rewarded under Aquino than under Marcos, despite occasional disappointments experienced under Aquino.

The Philippine experience also shows that the technocratic position is not monolithic. There were many open splits among leading technocrats (e.g. between Monsod and Ongpin) in policy matters during the Aquino regime which reflected basic differences in value orientation. Underlying the ostensible value neutrality of the leading technocrats, there was a strong commitment to conservative developmentalist models advocating the free-market philosophy, with emphasis on export-oriented industrialization and the WB/IMF propagated "adjustment" of the economy. On the other hand, a minority of technocrats, while not denying economic liberalism, showed a preference for more "nationalist" and "people-" or "cause-oriented" ideas of development. Their economic strategies focused on rural areas and employment generation and were aimed at improving the lot of ordinary Filipinos.

As defined in this work, technocracy is treated as essentially an autonomous activity, distinguished from ordinary bureaucratic practices. It enjoys discretion whereas bureaucracy is subject to the political will. It has been noted that at this stage it is impossible to frustrate the tendency of an autonomous "technocracy" to develop, and to restore traditional bureaucratic practices subject to political direction. Therefore it is necessary to bring technocracy under public control and make it more responsive to public interest. Several strategies have been explored to bring this about. One proposal has been to keep technocratic power in check by countervailing forces. Another proposal follows Yehezkel Dror's idea of special education for policy-makers in the art of policy-making and analysis. This involves the fostering of modern leaders who are able to deal with policy issues from a broad perspective, including both rational and extra-rational or value aspects. A similar proposal highlights the need to make political and educational elites and intellectuals alert to the danger of values dominated by industrial and technology-based systems and to assume leadership in propagating values that show concern for "a better quality of life." The principal proposal of this work aims at changing the technocratic concept in a more "humanist" direction by a modification of the methodology of technocratic thought, which, as the failure of technocratic planners to respond to "cause-oriented" issues and problems of poverty indicates, reflects essentially counter-humanist values. The new orientation seeks to integrate "humanist" objectives into policy designs prepared by technocratic planners, reflecting not only a commitment to technology and scientific rationality but, more importantly, a concern with issues of "humanist" interest and a better quality of life for ordinary people.

Notes

CHAPTER I

1. Principal sources on technocracy used in Chapter I:

David Apter, "The Role of the New Scientific Elite and Scientific Ideology in Modernization," in Sal Restivo and Christopher Vanderpool, eds., *Comparative Studies in Science and Society* (Columbus, Ohio: Charles E. Merrill Publishing Co., 1974).

Victoria M. Arcega, *Technocrats as Middlemen and Their Networks in the Philippine Rice Project: The Case of the Masagana 99* (Ph.D. Dissertation, Michigan State University, 1976).

______________, "Technocrats in the Philippine Rice Program: Middlemen in Masagana 99," *Philippine Journal of Public Administration*, April 1976.

Daniel Bell, *The Coming of Post-Industrial Society: A Venture in Social Forecasting* (New York: Basic Books, 1973).

James Burnham, *The Managerial Revolution* (New York: The John Day Co., 1941).

Aurelio B. Calderon, "American-Trained Technocrats: Preliminary View and Trends," *Philippine Journal of American Studies*, April 1979.

Hans Peter Dreitzel, "Social Science and the Problem of Rationality: Notes on the Sociology of Technocrats," *Politics and Society*, 11, 2 (Winter 1972).

Henry Elsner, *The Technocrats: Prophets of Automation* (New York: Syracuse University Press, 1967).

John Kenneth Galbraith, *The New Industrial State* (Penguin Books, 1967).

Robert Gilpin and Christopher Wright, eds., *Scientists and National Policy Making* (New York: Columbia University Press, 1964).

Joseph Haberer, *Politics and the Community of Science* (New York: D. Van Nostrand Reinhold Co., 1969).

Glenn Anthony May, *Social Engineering in the Philippines: The Aims, Execution, and Impact of American Colonial Policy, 1900–1913* (Westport, Connecticut: Greenwood Press, 1980).

Jean Meynaud, *Technocracy* (New York: The Free Press, 1969).

Frank Musgrove, *Ecstasy and Holiness* (London: Methuen and Co., 1974).

Romeo Ocampo, " Technocrats and Planning: Sketch and Exploration," *Philippine Journal of Public Administration*, January 1971.

Don Price, *The Scientific Estate* (London: Oxford University Press, 1965).

Sal P. Restivo and Christopher Vanderpool, *Comparative Studies in Science and Society* (Columbus, Ohio: Charles E. Merrill Publishing Co., 1974).

F. F. Ridley, " French Technocracy and Comparative Government," *Political Studies*, February 1966.

Theodore Rozsak, *The Making of a Counter-Culture: Reflections on the Technocratic Society and its Useful Opposition* (New York: Anchor Books, 1968).

Peter Self, *Econocrats and the Policy Process: The Politics and Philosophy of Cost-Benefit Analysis* (London: The Macmillan Press, 1975).

____________, *Administrative Theories and Politics* (London: George Allen and Unwin, 1972).

Technocracy Study Course (New York: Technocracy, Inc., 1934).

Allain Touraine, *The Post-Industrial Society* (New York: Random House, 1971).

2. Galbraith, *op. cit.*, p. 392.

3. *Ibid.*, p. 18.

4. *Technocracy, vii.*

5. Ellill, quoted in Frank Musgrove, *op.cit.*, p. 43.

6. Bell, *op. cit.*, p. 487.

7. *Ibid.*, p. 20.

8. Galbraith, *op. cit.*, p. 357.

9. *Ibid.*, p. 399.

10. *Ibid.*, p. 380.

11. *Ibid.*, p. 19.

12. *Ibid.*, p. 378.

13. Touraine, *op. cit.*

14. *Ibid.*, p. 49.

15. *Ibid.*, p. 98.

16. May, *op. cit.*, p. xv.

17. Arcega, *Technocrats and Middlemen*, p. 253.

18. *Ibid.*, p. 254.

19. *Ibid.*, p. 256.

20. Roszak, *op. cit.*, p. 7.

21. Self, *Administrative Theories*, p. 195.

22. Meynaud, *op. cit.*, p. 208.

23. Touraine, *op. cit.*, pp. 49–50.

24. Dreitzel, *op. cit.*, p. 170.

25. For an interesting illustration of this see "Acrobatic technocrats star in an Indonesian balancing act," *Far Eastern Economic Review*, May 16, 1980. The orientation of government technocrats is contrasted with the orientation of engineers filling posts in some top economic ministries. While the technocrats are said to show preference for rational macro (or integrated) economic systems which are expected to "assure steady, even-handed development," the engineers are said to have a micro-economic bias and to exude an essentially entrepreneurial spirit, i.e., to be action-oriented and prepared to take short cuts to get quick results.

26. Roszak, *op. cit.*, p. 8.

27. Arcega, *Technocrats and Middlemen*, p. 258.

28. Bell, *op. cit.*, p. 349.

29. W.W. Rostow, *The Stages of Economic Growth* (Cambridge University Press, 1960), pp. 12–13 and p. 1.

CHAPTER II

1. See, e.g., Renato Constantino, and Letizia R. Constantino, *The Philippines: The Continuing Past* (Quezon City: The Foundation for Nationalist Studies, 1978), Ch. 4.

2. Lucian W. Pye, *Aspects of Political Development* (Boston: Little, Brown and Co., 1966).

3. For Manuel Roxas' Inaugural Address of May 28, 1946, see the *Official Gazette*, Manila, May 1946, Vol. 42, No. 5, pp. 980–89, and T.A. Agoncillo and M.C. Guerrero, *History of the Filipino People* (Quezon City: Garcia Publishing, 1976) on the presidency of Manuel Roxas.

4. See, e.g., Agoncillo, *op. cit.*, on the presidency of Elpidio Quirino; also Quirino's speeches and his *The New Philippine Ideology* (Manila: Bureau of Printing, 1949).

5. Three principal economic evils were identified by the Bell Mission: low productive efficiency and low diversification of the economy, inequalities in the level of income of the people, and inefficiency and corruption in the government. The Mission's report included proposals for economic and social reforms, e.g., revision of tax structures, encouragement of investment in productive enterprises, diversification of production, improvement of agricultural production, a reduction of "undue" demands for imports, improvement of administration "to ensure honesty and efficiency in the government," as well as "an adequate program" of health and education on the side of social reforms.

6. Some of the agencies established during his Administration were: the Philippine Council for U.S. Aid (PHILCUSA); the President's Action Committee on Social Amelioration (PACSA); the Labor-Management Advisory Council; the Agricultural Credit Cooperatives Financing Administration (ACCFA), the Rural Bank, etc.

7. "The State of the Nation Address," January 23, 1956, in Agoncillo and Guerrero, *op. cit.*, pp. 550–52.

8. Constantino and Constantino, *op. cit.*, Ch. IX.

9. Agoncillo, *op. cit.*, on the presidency of Carlos Garcia.

10. Frank H. Golay, *The Philippines: Public Policy and Economic Development* (Ithaca, N.Y.: Cornell University Press, 1963), pp. 92, 95–97.

11. Agoncillo, *op. cit.*, p. 549.

12. *Ibid.*

13. *The Civil Service Recorder*, November 1957, p. 8.

14. D.A. Aytona (Commissioner of the Budget), "Highlights of The Five-Year Fiscal Plan (1957–1961)"; quotations on p. 165 and p. 173 respectively.

15. *Ibid.*, p. 165.

16. See, e.g., Diosdado Macapagal's addresses: " New Hope for the Common Man," " Our Mission under the New Era," " Win Back the People's Confidence" and " An Unfinished Revolution."

17. " Primer on President Macapagal's Socio-Economic Program" in *The Civil Service Recorder*, August 1962, p. 4.

18. *Ibid.*

19. *Ibid.*, p. 7.

20. Diosdado Macapagal, " Our Mission under the New Era" (Manila: National Media Production Center, 1962).

21. Armand Fabella, " The PIA and The First Year of the Socio-Economic Program," *The Fookien Times Philippine Yearbook 1963*, p. 113.

22. F. E. Marcos, " Mandate for Greatness." Address on the State of the Nation, January 24, 1966 (Manila: Bureau of Printing, 1966).

23. *Ibid.*, p. 4.

24. *Ibid.*, p. 6.

25. *Ibid.*, p. 7.

26. *Ibid.*

27. *Ibid.*, p. 17.

28. " Human Dignity—The True Revolution." Address, September 15, 1966 (Manila: Bureau of Printing, n.d.).

29. *Ibid.*, p. 11.

30. Jean Grossholtz, *Politics in the Philippines* (Boston: Little, Brown and Co., 1964); quotations on p. 73 and p. 74, respectively.

31. Golay, *op. cit.*, p. 415.

32. *Ibid.*, p. 11.

33. For illustrations, see, e.g. Abueva and de Guzman, eds., *Foundations and Dynamics of Filipino Government and Politics* (Quezon City: Bookmark, 1969).

34. S. N. Eisenstadt, " Bureaucracy, Bureaucratization and Debureaucratization," *Administrative Science Quarterly*, 4 (1959).

35. Grossholtz, *op. cit.*, p. 24.

36. Attempts at administrative reorganization were made by Quirino in 1949, Magsaysay in 1957, Garcia in 1958 and Macapagal in 1962.

37. In the words of the Dean of the Institute Carlos P. Ramos, " to provide present and prospective public servants with the education and skills that will prepare them for effective and responsible performance in their jobs." See " Program and Goals of the Institute of Public Administration, University of the Philippines," *Philippine Journal of Public Administration*, January 1957.

38. One instance of this recognition is the appearance in the prestigious *Fookien Times Yearbook* of a new section on " Law and Public Administration," which was to remain a permanent feature of this publication (with the exception of the two initial years of the Martial Law Administration, 1973 and 1974).

39. A. del Rosario, *The Civil Service Reporter*, April 1957, p. 8.

40. See Jose V. Abueva, ed., *Perspectives in Government Reorganization* (Manila: College of Public Administration, University of the Philippines, 1969).

41. Golay, *op. cit.*, p. 8.

42. *Ibid.*

43. Whether explained in the context of dyadic social relationships (Lande) or by an organic-hierarchical model of traditional authority (Agpalo), the government seems to have been expected to exhibit a paternalistic style in its

treatment of the people. Even modernization had to come to terms with this traditional element. As Golay put it in 1961 in the new vocabulary of modernization, " Filipinos . . . look upon government as a means to national integration and accelerated economic growth." (*Ibid.*, pp. 343–44).

44. *Ibid.*, pp. vii–viii.

45. *Ibid.*, p. 183.

46. The early plans were aimed mainly at providing incentives and protection to the fledgling manufacturing sector during the period of the policy of import substitution, lasting from 1949 to 1962. By the 1960s, however, the scope of planning had broadened and included an emphasis on socio-economic development and such areas of public interest as education, skills training, health and agricultural extension. (See, e.g, Corpuz, *The Philippines*, p. 12.) A major initiative in national planning was an ambitious land reform program passed under President Macapagal in 1963. The idea of " integrated" planning originated also during Macapagal's presidency.

47. Onofre D. Corpuz, *The Philippines* (Englewood Cliffs, N. J.: Prentice-Hall, 1964); Jean Grossholtz, *Politics in the Philippines* (Boston: Little, Brown and Co., 1964).

48. Corpuz, *ibid.*, p. 5.

49. *Ibid.*, p. 138.

50. *Ibid.*

51. *Ibid.*, p. 126.

52. *Ibid.*, p. 92.

53. Grossholtz, *op. cit.*, p. 79.

54. *Ibid.*, p. 107.

55. *Ibid.*, p. 79.

56. *Ibid.*, p. 107.

57. *Ibid.*, p. 49.

58. *Ibid.*, p. 78.

59. *Ibid.*, p. 73.

60. *Ibid.*, p. 75.

61. Raul P. de Guzman, " Politics in the Philippines," *Philippine Journal of Public Administration*, July–October 1971, p. 233.

62. *Ibid.*

63. *Ibid.*, p. 236.

64. For this theory see, e.g., S.N. Eisenstadt, *Modernization, Protest and Change* (Englewood Cliffs, N.J.: Prentice-Hall, 1966).

65. de Guzman, " Politics in the Philippines," p. 233.

66. F. E. Marcos, " Financial Management and Economic Development," a speech given on January 30, 1969 (no reference).

67. *Ibid.*, p. 63. It may be added that after 1972 the failings of Filipino democracy were described by Marcos in much more critical, indeed, black-and-white terms since the ideological interest of the " New Society" created a need to perceive the " Old Society" as a great evil that could have been overcome only by a new political arrangement.

68. Abelardo G. Samonte, " Reorganizing the Government for Development," in J. J. Carroll, ed., *The Philippine Economy in the 1970's: Prospects and Problems of Development* (Quezon City: University of the Philippines, 1972).

69. *Ibid.*, p. 28.

70. Jose V. Abueva, " Issues of Reorganization Strategy: The Government Survey and Reorganization Commission (1954–1956) and the Commission on Reorganization (1959–　　)" in Abueva, *op. cit.*, p. 9.

71. Romualdo B. Tadena, " Evaluating and Implementing Reorganization," *ibid.*, p. 85.

72. See Gerardo P. Sicat, *Economic Policy and Philippine Development* (Quezon City: University of The Philippines Press, 1972).

73. This argument is developed by Sicat, *ibid.*, and is later repeated in the " Ranis Report" published in 1974.

74. Gerardo P. Sicat, " Incentives Policy for Industrialization" in *The Philippine Economy in the 1970's,* p. 2.

75. " Magna Carta of Social Justice and Economic Freedom" (1969) by Speaker J.B. Laurel, Jr. (no reference), p. 17.

76. Romeo B. Ocampo, " Social Justice: An Essay on Philippine Social Ideology," *Philippine Journal of Public Administration*, July–October 1971, p. 290.

77. O.D. Corpuz, " Cultural Foundations" in Abueva *et al.*, *Foundations and Dynamics of Filipino Government and Politics* (Manila: Bookmark, 1969), p. 11.

78. Robert B. Stauffer, " A Legislative Model of Political Development, *Philippine Journal of Public Administration*, January 1967, p. 12.

79. de Guzman, " Politics in the Philippines," p. 238.

80. *Ibid.*

81. J.V. Abueva, ed., *Filipino Politics, Nationalism and Emerging Ideologies: Background for Constitution-Making* (Manila: Modern Book Co., 1975), p. 269.

82. *Ibid.*, p. 270.

83. Corpuz, " Cultural Foundations," pp. 16–17.

84. Aurelio B. Calderon, " American-Trained Technocrats: Preliminary Views and Trends," *Philippine Journal of American Studies*, April 1979, pp. 20–21.

85. Rafael P. de Tagle, " The New Era's Program Implementation Agency," *Chronicle Magazine*, June 8, 1963, p. 8.

86. William J. Pomeroy, *An American-Made Tragedy* (New York: International Publishers, 1974), p. 63.

87. Sixto K. Roxas Jr., "The Five-Year Development Program: A Summary Review of the Initial Phase," *The Fookien Times Yearbook 1963.*

88. _______________, "The Problem of Public Administration for Economic Development," *Philippine Journal of Public Administration*, January 1965.

89. *Ibid.*, p. 4.

90. *Ibid.*, p. 5.

91. Romeo B. Ocampo, "Technocrats and Planning: Sketch and Exploration," *Philippine Journal of Public Administration*, January 1971, p. 44.

92. Roxas, "The Five-Year Development Program," p. 118.

93. _______________, "Lessons from Philippine Experience in Development Planning." Paper presented at the Conference on Economic Planning in Southeast Asia in Honolulu, February 1–5, 1965, pp. 16–17.

94. Armand V. Fabella, "Some Aspects of the Strategy of Development Planning," in G.P. Sicat, ed., *The Philippine Economy in the 1960's* (Quezon City: University of the Philippines, 1964), pp. 56–57.

95. Ocampo, "Technocrats and Planning," p. 55.

96. See the relevant item in *The Civil Service Reporter*, October 1969.

97. For description of this particular outlook, see Roman Dubsky, "Max Weber and Development Administration: Conflicts and Identities in Administrative Thought—Toward a New Development Administration." Occasional Paper No. 72-9, College of Public Administration, University of the Philippines, October 1979.

98. Edward W. Weidner, "Development Administration: A New Focus for Research" in F. Heady and S.L. Stokes, eds., *Comparative Public Administration* (Ann Arbor: The University of Michigan, 1960), p. 98.

99. This definition is based on Bernard Schaffer, *The Administrative Factor* (London: Frank Cass, 1973), p. 245.

100. Weidner, *op. cit.*, p. 421.

101. Rafael M. Salas, " Reorganizing the Government," *The Fookien Times Yearbook 1966*, p. 188.

CHAPTER III

1. Alan Robson, " Martial Law in the Philippines and Its Aftermath" (Suva, Fiji: University of the South Pacific), SSED Working Paper No. 6, 1987.

2. Declaration of Martial Law on 21 September 1972.

3. Robson, *op. cit.*, p. 4.

4. Quotation from F. E. Marcos, *The Fookien Times Yearbook 1969*, p. 63.

5. Samuel P. Huntington, *Political Order in Changing Societies* (New Haven: Yale University Press, 1968).

6. Perfecto V. Fernandez, " From *Javellana* to *Sanidad*: An Odyssey in Constitutional Experimentation," in J. L. Cabildo, ed., *1976 Amendments and the New Constitution* (Quezon City: U.P. Law Center, Bookman Printing House, 1978), p. 38.

7. Clinton L. Rossiter, *Constitutional Dictatorship: Crisis Government in Modern Democracies* (Princeton: Princeton University Press, 1948).

8. Perfecto V. Fernandez, " Civic Liberties under Martial Law" in V.V. Mendoza, ed., *The Supreme Court under the New Constitution* (Manila: U.P. Law Center, 1977), p. 9.

9. Arturo M. Tolentino, " Explain type of government here," *Bulletin Today*, September 2, 1979, p. 1.

10. See Raul P. de Guzman, *et al.*, " Citizen Participation and Decision Making under Martial Law," *Philippine Journal of Public Administration*, January 1977, p. 5.

11. Ma. Aurora A. Carbonell, " Assessment of New Participative Mechanism," *Philippine Political Science Journal*, June–December 1977, p. 59.

12. F.V. Oamar and P. Rivera, " Local Government Development Since 1966," *Philippine Journal of Public Administration*, January–April 1975, pp. 11 and 14.

13. Belinda Aquino, *Dimensions of Decentralization and Development in the Philippines* (Ph.D. Dissertation, Cornell University, 1974), p. 286.

14. R.P. de Guzman and A.G. Pacho, " Control Mechanisms in Central-Local Government Relations," *Philippine Journal of Public Administration*, January–April 1975, p. 62.

15. For the sake of confidentiality the name of the author must be kept anonymous.

16. *Ibid.*

17. *Ibid.*

18. Sa-Myung Park, *The State, Revolution and Development* (Ph.D. Dissertation, State University of New York at Buffalo, May 1988), p. 253.

19. *Ibid.*, pp. 255–56.

20. *Ibid.*, p. 257.

21. *Ibid.*, pp. 257–59.

22. See F. E. Marcos, " The Practice of Good Government," *The Civil Service Reporter*, p. 10.

23. F.E. Marcos, *The Practice of Government* (Department of Public Information). Address delivered at the Symposium-Seminar on Good Government, Quezon City, November 24, 1975, p. 3.

24. F. E. Marcos, *A Dialogue with My People: Selected Speeches of Ferdinand E. Marcos*, ed. Francisco S. Tatad (Manila: Department of Public Information, 1973), p. 190.

25. Excerpt from *Five Years of the New Society, Philippine Daily Express*, March 17, 1978, p. 6.

26. Joseph P. McCallus, "The Myths of the New Filipino: Philippine Government Propaganda During the Early Years of Martial Law," *Philippine Quarterly of Culture and Society*, 17 (1989), p. 129.

27. Letter of Instruction No. 614. See "Leaders and Leadership Training," *The Civil Service Reporter*, November 1977, p. 11. The preceding quotation from Marcos is found in "Salient points of speech on martial law," *Philippine Daily Express*, June 13, 1978, p. 17.

28. See, e.g., "Hundreds of officials on way out," *Bulletin Today*, February 16, 1979, pp. 1, 12; or "Purge pushed; other local executives out soon," *The Times Journal*, June 21, 1978, pp. 1, 8.

29. See, e.g., "Marcos orders cuts in expenses," *Philippine Daily Express*, February 10, 1979, pp. 1, 6.

30. See, e.g., "Will penalize any delay in processing of papers," *Bulletin Today*, February 11, 1979, pp. 1, 10; "Government 'merit' boards up," *Bulletin Today*, February 16, 1979, pp. 1, 12.

31. "Back to early '70s" (editorial), *Philippine Daily Express*, August 10, 1979, p. 4.

32. See Fred W. Riggs, "The Sala Model: An Ecological Approach to the Study of Comparative Administration," *Philippine Journal of Public Administration*, January 1962, pp. 414–17.

33. See Roman Dubsky, "The Government's Strategy for Small Industries and Intensive Use of Labor in the Philippines" (Unpublished paper), 1976, pp. 17, 29.

34. Reference is made to the Investment Incentives Act.

35. "Industry Rationalization," *Journal of the American Chamber of Commerce in the Philippines*, June 1971, p. 3.

36. Romeo Ocampo, "Technocrats and Planning: Sketch and Exploration," *Philippine Journal of Public Administration*, January 1971.

37. *Sharing in Development, A Programme of Employment, Equity, and Growth for the Philippines* (popularly known as the "Ranis Report") (International Labor Office, Geneva, 1974).

38. For summary and recommendations, see Chs. 2 and 12; for the "policy package," pp. 35–51.

39. From *Five Years of the New Society, Philippine Daily Express*, March 10, 1978, p. 8.

40. *Ibid.*

41. Gerardo P. Sicat, *The New Economic Directions in the Philippines* (Manila: National Economic Development Authority, 1974), p. 121.

42. F. E. Marcos, *Notes on the New Society of the Philippines*, 3rd edition (Manila: Marcos Foundation, 1974), p. 10.

43. "Business and Government." Speech of Marcos on November 23, 1973 in Marcos, *A Dialogue With My People.*

44. "A Period of Trial." Address of Marcos at the National Convention of Manufacturers and Producers, August 16, 1974.

45. "F.M. throws out bill on nationalization," *Bulletin Today*, December 8, 1978, p. 13.

46. Sicat, *The New Economic Directions*, p. 244.

47. See *Philippine Progress* (a BOI journal), Vol. X, 4th quarter, 1975, Table 1, p. 11. Data indicate that between 1968 and 1975 the BOI approved a total of P 3.6 billion foreign investments, which constituted 61 percent of total BOI-sponsored investments.

48. *The Five Year Philippine Development Plan, 1978–1982* (Manila: The Republic of the Philippines, September 21, 1977).

49. See "Sicat unveils policy on jobs," *Bulletin Today*, October 13, 1980, p. 14, and "Marcos submits 54.8-B outlay," *Bulletin Today*, July 29, 1980, pp. 1, 11.

50. See, e.g., "IBP studies P 34-B budget," *Philippine Daily Express*, June 14, 1978.

51. *The Five Year Philippine Development Plan, 1978-1982*, p. 355.

52. See, e.g., Agustin Kintanar Jr., Monograph No. 17 of the University of the Philippines Press (1976), and Honesto Mendoza and Leonor Magtolis-Briones, "Post-Martial Law Developments in Philippine Fiscal Administration" (1976). See also *U.P. Newsletter*, July 19 and October 25, 1976 respectively for references to these studies.

53. Sicat, *The New Economic Directions*, pp. 240, 242.

54. From *Five Years of the New Society*, *Philippine Daily Express*, March 7, 1978, p. 7.

55. These events are extensively described in Robin Broad's *Unequal Alliance, 1979–1986* (Quezon City: Ateneo de Manila University Press, 1988).

56. Robson, *op. cit.*, particularly sections on "The Economic Context of Martial Law" and "The Coup de Grace." Also in G.P. Sicat's "A Historical and Current Perspective of Philippine Economic Problems." An address delivered at the Meeting of the Philippine Economic Society, December 8, 1984. Read particularly pp. 42–60.

57. "Development Policy and Planning: The Philippine Experience" (Manila: CPA Research Team, University of the Philippines, 1973), p. 41.

58. On the BOI approach to social policy, see Ocampo, "Technocrats and Planning," *op. cit.*

59. *Ibid.*, p. 53.

60. The plan (of 1973) declared that "the principle of horizontal integration, wherever this does not pose dis-economies in the development of a sector, should be preferred to vertical integration since the horizontal pattern allows more small and medium scale industries to participate in the development of industrial sector."

61. The formula for labor intensiveness was expressed as the ratio of imported capital assets to actual profit in employment in each year of full operation. See this writer's paper on "Small Industries and Intensive Use of Labor," *op. cit.*

62. E.g., four criteria have come to be used in computing the worthiness of projects for government loans, which suggests a departure from the traditional narrow economics-oriented criteria.

63. From *Five Years of the New Society*, in *Philippine Daily Express*, March 7, 1978, p. 7.

64. *The Five Year Philippine Development Plan, 1978–1982*, Table 17.3, p. 368.

65. *Ibid.*, see Part 3, Chs. 9–12.

66. *Ibid.*, Table 9.1.

67. "FM: 'Emphasis on common man'," *Philippine Daily Express*, May 7, 1979, p. 1.

68. F. E. Marcos, *Today's Revolution: Democracy* (Manila: Ferdinand E. Marcos, 1971), p. 136.

69. PACRIS, *Marcos and Humanism* (Quezon City: National Printing Co., 1977), p. 35.

70. *Ibid.*

71. "FM: 'Emphasis on common man'," p. 1.

72. "Institutionalizing humanist ideology" (editorial), *Sunday Express*, June 4, 1978, p. 4.

73. "FL outlines objectives of new ministry," *Bulletin Today*, June 10, 1978, pp. 1, 2; also "Human settlements for R.P.," *The Times Journal*, June 2, 1978, p. 8.

74. *Ibid.*, "FL outlines objectives," p. 12.

75. See "Focus: Human Settlements," *Philippine Planning Journal*, October 1975, particularly pp. 1–9.

76. "BLISS project to upgrade slums," *Bulletin Today*, September 11, 1979, p. 27.

77. "FL inaugurates BLISS projects in three areas," *Bulletin Today*, June 12, 1979, p. 8.

78. "Traders' group responds to 'Big Brother' project," *Philippine Daily Express*, January 4, 1979, p. 2.

CHAPTER IV

1. See *Forestry Administrative Order No. 11* (*Revised*) (Republic of the Philippines, Department of Natural Resources, Bureau of Forest Development, Diliman, Quezon City: no date).

2. See Owen J. Lynch Jr. and Kirk Talbott, "Legal Responses to the Philippine Deforestation Crisis," *Journal of International Law and Politics*, pp. 682–86.

3. "Kaingin management" is defined as referring to "a portion of the forest land which is subjected to shifting and/or permanent slash-and-burn cultivation." Definitions, Sect. 1 (p), P.D. 1559.

4. Richard Dorall, "Comments on proposed Ministry of Energy regulations for the integrated management of resources in watershed reservations" (unpublished paper), 1979, p. 5.

5. For the text of the proposed bill, see *Bulletin Today* (Sunday), January 28, 1979.

6. " Law enforcement" (editorial), *Bulletin Today*, January 25, 1979, p. 6.

7. Lynch and Talbott, *op. cit.*, pp. 686–90.

8. The main sources about the two projects (including reports, resolutions, correspondence, etc.):

Montanosa Social Action Center, " The Chico River Basin Development Project: A Case Study on Development" (A Revised Summary), Baguio City, 1977 (?) (Mimeo).

Richard Dorall, " The Tinggians of Abra and Cellophil: A Situation Report" (revised edition). Prepared for the 2nd National Conference of the Ugnayang Pang-Aghamtao (UGAT), Inc., (Anthropological Association of the Philippines) in Baguio City, April 22–24, 1979.

Other sources:

Richard Dorall, "The Case Against Cellophil," *Who*, January 13, 1979.

Forestry Administrative Order No. 11 (Revised), *The Official Gazette*, September 14, 1970.

Noel C. Villalba, "Dark Clouds Over the Cordilleras," *Who*, October 13, 1979.

Max Garming, " What price growth? (The Case of the Kalingas of Northern Luzon)," *The La Sallian*, July 1979.

Albert Lee, " Tinggians have forged a peace pact," *Panorama*, March 25, 1979.

" Cellophil Resources Corporation: Its Concern and Commitments," *Focus*, June 30, 1979.

Daniel Llanto, "An industrial firm with civic conscience," *Focus*, June 30, 1979.

Francisco Suling, " Is a Giant Stamping out the Hill-Tribes?," *Who*, September 15, 1979.

9. *The Rangtay* (CRC's official newsletter), October 1977.

10. See Dorall, " The Tinggians," p. 6.

11. Daniel E. Llanto, *op. cit.*

12. See *The Northern Star*, a Baguio weekly, April 29, 1979.

13. " On the Chico River Project: The Progress of the People . . . Nozzle-End of a Gun, *Impact*, pp. 214–15.

14. See Montanosa, *op. cit.*, p. 16.

15. " Bishop clarifies dam stand," *Bulletin Today*, August 16, 1980, p. 10.

16. See Montanosa, *op. cit.*, pp. 24–25.

17. *Ibid.*, p. 42 for the letter to the President and p. 43 for the pastoral letter of January 20, 1977.

18. *Ibid.*, p. 37.

19. " NPC President meets with Kalinga priests on dam issue," *Montanews*, May–June 1979, p. 3. Perhaps not an untypical description of how the people felt about the government came, in early 1979, from Max Garming, a political scientist and a native Kalinga:

> " At present, the mountain people are in a suicidal (mood) in their opposition to the dam. Because of repeated bitter experiences, mistrust of government moves is deeply internalized among them; they will give no ear to reason. Yet for this, the government has only itself to blame." (*Who*, February 10, 1979).

20. " Bishop refutes Enrile's accusations against K-A [Kalinga Apayao] religious," *Bulletin Today*, June 14, 1979, p. 3.

21. Dorall, " The Tinggians," p.4.

22. (Manila, September 1966).

23. (National Economic Council, June 26, 1970).

24. (Republic of the Philippines, July 23, 1971).

25. Summary of the Four-Year Development Plan, FY 1972–75 in *The Philippine Economy in the 1970's, Prospects and Problems of Development* (Institute of Economic Development and Research, University of the Philippines, 1972), p. 255.

26. *Ibid.*

27. " Strength through Crisis, Growth in Freedom." Message delivered by President Marcos on January 24, 1972 (Manila: Bureau of Printing, 1972).

28. (Manila: the Republic of the Philippines, 1973).

29. (Manila: the Republic of the Philippines, September 21, 1977).

30. See *Ilocos (Regional) Five-Year Development Plan 1978–82* (Quezon City: NEDA Production Unit, 1977), pp. 110–11.

CHAPTER V

1. F. E. Marcos, *Today's Revolution: Democracy* (Manila, 1971), p. 50.

2. In *Philippine Daily Express*, March 10, 1978, p. 8.

3. *Ibid.*

4. *Ibid.*

5. Jose Crisol, " Building the New Society," *The Civil Service Reporter*, December 1972, p. 5.

6. " Professionals must be emulated—FL," *Bulletin Today*, June 27, 1979; quotations nos. 1 and 3 on p. 10, no. 2 on p. 2.

7. F. E. Marcos, *A Dialogue with My People: Selected Speeches of Ferdinand E. Marcos*, p. 210.

8. " Total Experience," editorial, *The Civil Service Reporter*, September–October 1974, p. 8.

9. " Pres. Marcos Inaugurates Career Executive Service," *ibid.*

10. " FM eyes non-government man for BIR post," *The Times Journal*, October 16, 1980, p. 1.

11. F.E. Marcos, *Towards a Filipino Ideology* (Manila: World Publishing, 1979).

12. *Ibid.*, p. 4.

13. *Ibid.*, p. 46.

14. See Raul P. de Guzman *et al.*, " Citizen Participation and Decision Making Under Martial Law," *Philippine Journal of Public Administration*, January 1977, p. 14.

15. F. E. Marcos, " Education and the Realities of National Life," *Philippine Development*, January 15, 1977, p. 8.

16. " Corpus vows to make key institutions move—and fast," *Bulletin Today*, January 5, 1981, p. 28.

17. " Filipinos Push Education Reforms," *The Philippines Today* (Department of Public Information), April 1976, p. 72.

18. See *Four-Year Development Plan, FY 1974–77 Extract* (NEDA, Manila, 1973), pp. 21, 22; *Five-Year Philippine Development Plan, 1978–1982*, p. xxii.

19. " Student councils, papers restored," *Bulletin Today*, January 27, 1981, pp. 1, 8; quotation on p. 8.

20. See Edward M. Masa's *The Higher Civil Service in the Philippines.* (Ph.D. Dissertation, University of the Philippines, March 1976).

CHAPTER VI

1. Edward I. M. Masa, *The Higher Civil Service in the Philippines* (Ph.D. Dissertation, University of the Philippines, 1976), p. 335.

2. A variety of works touches on the influence of technocratic planners and the role of the World Bank and the International Monetary Fund in the Philippine economy (e.g. by Robin Broad, Vivencio R. Jose, Walden Bello, Belinda Aquino). They are mostly critical of foreign (American, more recently Japanese) economic influence, promote economic nationalism and condemn the two international agencies and their Philippine technocratic allies for blindly following the global capitalist system.

3. See, e.g., " Development for Whom? The Industrialization of Barrio Nabacaan" (mimeo. unsigned, May 1975) or Kido Junko, " Kawasaki Steel Corporation's Sinter Plant in Mindanao" (mimeo); Lindy Washburn, " Our Lake for Others? the Maranao and the Agus River Hydroelectric Project," *Research Bulletin*, Vol. III, No. 3–4, November–December 1977.

4. Blondie Po, *Rural Organizations and Rural Development in the Philippines: A Documentary Study* (Quezon City: Ateneo de Manila, Institute of Philippine Culture, 1977), p. 128.

5. *Ibid.*, p. 129.

6. Mary Racelis Hollnsteiner, " Mobilizing the Rural Poor Through Community Organization," *Philippine Studies*, Vol.27/Third Quarter 1979, p. 403.

7. *Ibid.*, p. 389.

8. An excellent study on this question in the context of Nazi Germany is Ian Kershaw's *The Nazi Dictatorship: Problems and Perspectives of Interpretation* (Baltimore, Md., Arnold, 1985).

9. See Gerardo P. Sicat, " A Historical and Current Perspective of Philippine Economic Problems." Address delivered at the 21st Annual Meeting of the Philippine Economic Society, December 8, 1984, particularly pp. 51–54.

10. *Ibid.*, p. 60.

11. See Sa-Myung Park's *The State, Revolution and Development.* According to Park (p. 263), " of the top 50 corporations in terms of total sales in 1981, for instance, 11 corporations accounting for 33 percent of the total sales were controlled by 'crony capital,' corporations accounting for 25 percent of the total sales by local capital, 12 corporations accounting for 25 percent of the total sales by foreign capital, and the remainder by state capital. As a result, in the words of Noble, 'Party switching was replaced by defections; policy making, once subject to executive-legislative stalemate, was now hampered by bureaucratic rivalries'."

12. Lela Garner Noble, " Politics in the Marcos era," in John Bresman, ed., *Crisis in the Philippines: the Marcos Era and Beyond* (Princeton, N.J.: Princeton University Press, 1986), p. 96.

13. Robin Broad, *Unequal Alliance, 1979–1986* (Quezon City: Ateneo de Manila University Press, 1988), pp. 172–77.

14. *Ibid.*, quoted on p. 176.

15. The Philippine Economic Society, " Report on a Survey of Opinions of the Membership of the Philippine Economic Society on the Current Economic Crisis," *Philippine Economic Journal*, 24, (1985), pp. 64–79.

16. Mark Turner, " The Political Economy of the Philippines: Critical Perspectives," *Pacific Affairs*, 57(3), 1984, p. 465.

17. Noble, *op. cit.*, p. 102.

18. John T. Doherty, " Who Controls the Philippine Economy: Some Need Not Try as Hard as Others," in

Belinda A. Aquino, ed., *Cronies and Enemies: The Current Philippine Scene*, Philippine Studies Occasional Paper No. 5 (Honolulu: Philippine Studies Program, University of Hawaii, 1982).

CHAPTER VII

1. Blas F. Ople, " Politics of Stalemate" in F.E. Marcos *et al.*, *Toward the New Society* (National Media Production Center, 1974), p. 18.

2. *Technocracy Study Course*, (New York: Technocracy, Inc., 1974), pp. 211, 210.

3. Max Garming, " Footnote to the Kalinga Problem," *The Philippine Collegian*, September 24, 1975, p. 5.

4. Susanne J. Bodenheimer, " The Ideology of Developmentalism: American Political Science's Paradigm-Surrogate for Latin American Studies," *Berkeley Journal of Sociology*, Vol.XV (1970), p. 10.

5. Roman Dubsky, " Max Weber and Development Administration: Conflicts and Identities in Administrative Thought—Toward a New Development Administration," Occasional Paper No. 72-9 (October 1979), College of Public Administration, University of the Philippines, p. 10.

6. *Ibid.*, pp. 1–4. The technocratic type of organization is identified here with the concept of development administration.

7. Rolf E. Vente, *Planning Processes: The East African Case* (Munich: Weltforum Verlag, 1970), pp. 107–08, 97–98.

8. The first quotation by Emmette S. Redford in *The Municipal Year Book, 1974* (Washington, D.C., International City Management Association, 1974), pp. 393–97; the second quotation by I.L. Richardson and S. Baldwin in *Public Administration: Government in Action* (Columbus, Ohio, Charles E. Merrill, 1976), p. 13.

9. John K. Galbraith, *The New Industrial State* (Penguin Books, 1967), p. 393.

10. *Ibid.*, p. 380.

11. See K. Mannheim, *Freedom, Power, and Democratic Planning* (London, 1951).

12. Vente, *op. cit.*, p. 108.

13. Cf. Mannheim, p. 7.

14. Galbraith, *op. cit.*, p. 383.

15. *Ibid.*

16. David A. Buchanan and A.A. Huczynski, *Organizational Behavior* (London: Prentice-Hall International, 1985), pp. 474–76.

17. Yehezkel Dror, *Ventures in Policy Sciences* (New York: Elsevier, 1977), pp. 217–22; quotation on p. 218.

18. Dubsky, *op. cit.*, p. 15.

19. *Ibid.*

20. Brian Fay, *Social Theory and Political Practice* (London: George Allen and Unwin, 1975), p. 94.

21. Yehezkel Dror, *Design for Policy Sciences* (New York: Elsevier, 1978), Chs. 2, 9.

22. Bertram M. Gross, "Destructive Decision-Making in Developing Countries," *Policy Sciences* 5 (1974), p. 213.

CHAPTER VIII

1. Sandra Burton, "Aquino's Philippines: the Center Holds," *Foreign Affairs*, Vol. 65, No. 3, 1987, p. 534.

2. The Philippine Economic Society, "Report on a Survey of Opinions of the Membership of the Philippine Economic Society on the Current Economic Crisis," *Philippine Economic Journal*, 24, 1 (1985), pp. 64–79.

3. Asian Development Bank, " Economic Memorandum on the Philippines," 4 November 1986, p. 6.

4. More fully, the reorganization was to be based on five principles: promotion of private initiative, decentralization, cost effectiveness, frontline services efficiency and accountability. See " Gov't reorganization approved," *Business Day*, August 7, 1986, p. 1.

5. " Cory's fading magic," *Far Eastern Economic Review*, October 1, 1987, p. 10.

6. " Philippine Project Illustrates Policy Splits," *Asian Wall Street Journal*, March 12, 1990, p. 1.

7. " New helmsman, old course at Petron," *Far Eastern Economic Review*, July 7, 1988, p. 90.

8. *Ibid.*, p. 91.

9. " Cronyism, Communism Stall Philippine Progress," *Asian Wall Street Journal*, February 7, 1990.

10. *Ibid.*

11. " Avoiding action," *Far Eastern Economic Review*, January 11, 1990, p. 8, and " High flier," *Far Eastern Economic Review*, February 14, 1991, pp. 16, 17.

12. " Manila on the brink," *ibid.*, November 8, 1990, pp. 48, 49.

13. " The enemy is us," *ibid.*, p. 51.

14. " The boss system," *ibid.*, September 14, 1989, pp. 36 and 36–43.

15. " Still all in the family," *ibid.*, March 26, 1987, p. 70.

16. " Power to the patron," *ibid.*, January 17, 1991, p. 23.

17. *Ibid.*

18. " Strike two against Cory," *ibid.*, April 19, 1990, p. 11.

19. " The Philippines," *South Survey*, February 1989, p. 43.

20. Mark Turner, " The Philippines: disillusioned and disunited," *Asia-Pacific Defence Reporter*, July 1990, p. 12.

21. A.D.B., *op. cit.*, p. 5.

22. " A new strategy for industrial renaissance," *Far Eastern Economic Review*, November 6, 1986, p. 80.

23. " UP economists gain influence," *Business Day*, May 16, 1986, p. 2.

24. " Nod seen for NEDA scheme," *Business Day*, May 7, 1986, p. 2.

25. " A new strategy for industrial renaissance," p. 80.

26. " A gov't economic plan same as '83 program: Virata," *Business Day*, May 12, 1986, p. 2.

27. B.M. Villegas, " Mrs Aquino's Economic Game Plan," *Asian Wall Street Journal*, March 19, 1986.

28. " Aquino orders 'fire sale' of Government corporations," *The Australian Financial Review*, December 12, 1986, p. 11.

29. " Manila Cabinet Clears Investment Code," *Asian Wall Street Journal*, June 1987, p. 3.

30. " BOI to shed off regulatory orientation," *Manila Bulletin*, September 20, 1986, p. 17.

31. " Gov't readies P3B to pump-prime the economy," *Business Day*, July 7, 1986, p. 1.

32. " Gov't policies worry business," *Business Day*, July 1, 1986, p. 1; and " Policy agenda hit," *Business Day*, June 27, 1986, p. 2 (includes quotations).

33. " IMF drops insistence on austerity measures," *Business Day*, July 24, 1986, p. 1.

34. " Broad outline of gov't economic thrust bared," *Business Day*, September 3, 1986, p. 1.

35. *Ibid.*, p. 2.

36. " Gov't to lift all import curbs before yearend," *ibid.*, June 27, 1986, p. 1.

37. " Give Cory time," *Manila Bulletin*, March 23, 1986, p. 6.

38. " Arroyo assails Ongpin policies," *Manila Bulletin*, March 16, 1987, p. 1; and " Monsod Says Manila Will Shun Any New IMF Reform Proposals," *Asian Wall Street Journal*, October 1, 1988.

39. " Down to business," *Far Eastern Economic Review*, July 13, 1989, p. 68.

40. *Ibid.*

41. " Manila on the brink," *ibid.*, November 8, 1990, p. 49.

42. *Ibid.*

43. *Ibid.*

44. " Duty bound," *ibid.*, September 13, 1990, p. 42.

45. Editorial, *ibid.*

46. *Ibid.*

47. Solita Monsod, " Standing for Development," *The Fookien Times Philippines Yearbook 1989*, p. 276.

48. *Ibid.*

49. *Ibid.*

50. " Congress of words, not acts," *Far Eastern Economic Review*, July 12, 1990, p. 36.

51. Victor O. Ramos, " Developing Our Forests," *The Fookien Times Philippines Yearbook*, p. 158.

52. *Ibid.*

53. *Philippine Development Plan 1987–1992* (Manila: 1986), p. 116.

54. *D.P. 1987–1992*, Ch. 4.

55. Ramos, *op. cit.*, p. 333.

56. "The forest problem," *Manila Bulletin*, May 18, 1987, p. 6.

57. Owen J. Lynch Jr. and Kirk Talbott, "Legal Responses to the Philippine Deforestation Crises," *Journal of International Law and Politics* (New York University), Vol. 20, 1988, No. 3, p. 687.

58. *Ibid.*, p. 689.

59. *Ibid.*, p. 687.

60. *Ibid.*, pp. 690–93, 702–03.

61. *Ibid.*, p. 695.

62. *Ibid.*

63. "Entrenched concessions," *Far Eastern Economic Review*, February 25, 1988, p. 19.

64. Ramos, *op. cit.*

65. *DP 1987–1992*, p. 114.

66. Ramos, *op. cit.*, p. 333.

67. World Bank, "The Philippines: Issues and Policies in the Industrial Sector" (Vol. I), Report no. 6706-PH, July 30, 1987.

68. *Ibid.*, p. vi.

69. *Ibid.*, p. 74.

70. *Ibid.*, p. 72.

71. *DP 1987–1992*, p. 338.

72. World Bank, "Issues," *op. cit.*, pp. vi–vii.

73. *Ibid.*, p. 73.

74. *Ibid.*, p. 72.

75. *DP 1987–1992*, pp. 338–41.

76. Sa-Myung Park, *op. cit.*, p. 266.

77. "The Philippine paradox," *Far Eastern Economic Review*, July 12, 1990, p. 34.

78. "Servicing the new farm elite," *South*, August 1988, p. 32.

79. *Ibid.*

80. *Ibid.*

81. *Ibid.*

82. "Power to the plutocrats," *Far Eastern Economic Review*, July 12, 1990, p. 40.

83. "Poor get poorer," *ibid.*, August 18, 1988, p. 34.

84. *Ibid.*

Bibliography

BOOKS

Abueva, Jose V. *Perspectives in Government Reorganization.* Manila: College of Public Administration, University of the Philippines, 1969.

___________. *Filipino Politics, Nationalism and Emerging Ideologies: Background for Constitution-Making.* Manila: Modern Books Co., 1975.

___________ and Raul P. de Guzman, eds. *Foundations and Dynamics of Filipino Government and Politics.* Manila: Bookmark, 1969.

Agoncillo, Teodoro A. and Milagros C. Guerrero. *History of the Filipino People.* Quezon City: Garcia Publishing, 1970.

Aquino, Belinda. *Dimensions of Decentralization and Development in the Philippines.* Ph.D. Dissertation, Cornell University, 1974.

___________. *The Politics of Plunder: The Philippines Under Marcos.* Manila: Great Books Trading /UP College of Public Administration, 1987.

Arcega, Victoria M. *Technocrats as Middlemen and Their Networks in the Rice Project: The Case of Masagana 99.* Ph.D. Dissertation, Michigan State University, 1970.

Averch, H.A., F.H. Denton and J.E. Kochler, *A Crisis in Ambiguity; Political and Economic Development in the Philippines.* Sta. Monica, California: Rand Corporation, 1970.

__________. *The Matrix of Policy in the Philippines.* Princeton, N.J.: Princeton University Press, 1970.

Bell, Daniel. *The Coming of Post-Industrial Society: A Venture in Social Forecasting.* New York: Basic Books, 1973.

Bello, Walden et al. *Development Debacle: The World Bank in the Philippines.* San Francisco: Institute for Food and Development Policy, 1982.

Braibanti, Ralph J.D. (ed.). *Political and Administrative Development.* Durham, N.C.: Duke University Press, 1969.

Brillantes, Alex Bello. *Explaining Philippine Authoritarianism: Martial Law in 1972.* Ph. D. Dissertation, University of Hawaii, 1986.

Broad, Robin. *Unequal Alliance 1979–1986: The World Bank, the International Monetary Fund, and the Philippines.* Quezon City: Ateneo de Manila University Press, 1988.

Carroll, John J. (ed.). *Philippine Institutions.* Manila: Solidaridad Press, 1970.

__________. *Prospects and Problems of Development.* Institute of Economic Development and Research, University of the Philippines, 1972.

Chapman, William. *Inside the Philippine Revolution: The New People's Army and its Struggle for Power.* New York: W.W. Norton and Co., 1987.

Constantino, Renato and Letizia R. Constantino. *The Philippines: The Continuing Past.* Quezon City: The Foundation for Nationalist Studies, 1978.

Corpuz, Onofre D. *The Bureaucracy in the Philippines.* Manila: Institute of Public Administration, University of the Philippines, 1957.

__________. *The Philippines.* Englewood Cliffs, N.J: Prentice-Hall, Inc., 1970.

Development in the '70s. Fifth Annual Seminar for Student Leaders, Tagaytay, 1973.

Elsner, Henry. *The Technocrats: Prophets of Automation.* New York: Syracuse University Press, 1967.

Francisco, Gregorio Jr. *Higher Civil Servants in the Philippines.* Manila: College of Public Administration, University of the Philippines, 1959.

Galbraith, John Kenneth. *The New Industrial State.* Penguin Books, 1967.

Gilpin, Robert and Christopher Wright (eds.). *Scientists and National Policy-Making.* New York: Columbia University Press, 1964.

Golay, Frank H. *The Philippines: Public Policy and National Economic Development.* Ithaca, N.Y.: Cornell University Press, 1961.

Goodno, James B. *The Philippines: Land of Broken Promises.* London: Zed Books, 1991.

Grossholtz, Jean. *Politics in the Philippines.* Boston: Little, Brown and Co., 1964.

de Guzman, Raul P. (ed.). *Patterns in Decision-Making.* Manila: College of Public Administration, University of the Philippines, 1963.

____________ and Mila A. Reforma. *Government and Politics in the Philippines.* Singapore: Oxford University Press, 1988.

Haberer, Joseph. *Politics and the Community of Science.* New York: D. Van Nostrand Reinhold Co., 1969.

Hayden, Joseph. *The Philippines: A Study in National Development.* New York: The MacMillan Co., 1942.

Hawes Gary. *The Philippine State and the Marcos Regime: The Politics of Export.* Ithaca: Cornell University Press, 1987.

Lande, Carl. *Leaders, Factions, and Parties: The Structure of Philippine Politics.* New Haven: Yale University Press, 1965.

__________ (ed). *Rebuilding a Nation: Philippine Challenges and American Policy.* Washington, D.C.: Washington Institute Press Book, 1987.

Lallana, Emmanuel C. *Class, State and Crisis in the Philippines.* Ph.D. Dissertation, University of Hawaii, 1986.

Lichauco, A. *The Lichauco Paper: Imperialism in the Philippines.* New York: Monthly Review Press, 1973.

Macapagal, Diosdado P. *Blueprint for Progress: Primer on Macapagal's Five-Year Socio-Economic Program.* Presidential Committee on Administration Performance, Efficiency, n.d.

__________. *New Hope for the Common Man.* Manila: Bureau of Printing, 1962-1965.

Machado Kit. *Leadership and Organization in Philippine Local Politics.* Ph.D. Dissertation, University of Washington, 1972.

Majul, Cesar A. *The Political and Constitutional Ideas of the Philippine Revolution.* Quezon City: University of the Philippines Press, 1967.

Mangahas, Mahar. *Measuring Philippine Development.* Manila: Development Academy of the Philippines, 1976.

Marcos, Ferdinand E. *Development for the New Society.* Manila: Bureau of National and Foreign Information, Department of Public Information, 1974.

__________. *A Dialogue with My People: Selected Speeches of Ferdinand E. Marcos.* (ed. by F.S. Tatad) Manila: Department of Public Information, 1973.

__________. *Notes on the New Society.* Manila: National Media Production Center, 1973.

__________. *Notes on the New Society of the Philippines.* (3rd printing) Manila: Marcos Foundation, Inc., 1974.

__________. *The Democratic Revolution in the Philippines.* Englewood Cliffs, N.J.: Prentice-Hall International, 1974.

__________. *Today's Revolution: Democracy.* Manila, 1979.

__________. *Towards a Filipino Ideology.* Manila, 1979.

Masa, Edward M. *The Higher Civil Service in the Philippines.* Ph.D. Dissertation, University of the Philippines, March 1976.

Marcos, Imelda R. *Manila: Toward the City of Man.* Manila: National Media Production Center, 1976.

May, Ron and F. Nemenzo (eds). *The Philippines After Marcos.* Croom Helm, Sydney, 1985.

Meynaud, Jean. *Technocracy.* New York: The Free Press, 1969.

Milne, R.S. (ed.). *Planning for Progress.* Manila: University of the Philippines, 1960.

PACRIS, *Marcos and Humanism.* Quezon City: National Printing Co., 1977.

Park, Sa-Myung. *The State, Revolution and Development: A Comparative Study of Transformation of the State in Indonesia and the Philippines.* Ph.D. Dissertation, State University of New York at Buffalo, May 1988.

Power, John H. and Gerardo P. Sicat. *The Philippines: Industrialization and Trade Policies.* London: Oxford University Press, 1971.

Price, Don. *The Scientific Estate.* London: Oxford University Press, 1965.

Quirino, Elpidio. *The New Philippine Ideology.* Manila: Bureau of Printing, 1949.

Restivo, Sal P. and Christopher K. Vanderpool. *Comparative Studies in Science and Society.* Columbus, Ohio: Charles E. Merrill Publishing, 1974.

Rossiter, Clinton L. *Constitutional Dictatorship: Crisis Government in the Modern Democracies.* Princeton, N.J.: Princeton University Press, 1948.

Rozsak, Theodore. *The Making of Counter-Culture: Reflections on the Technocratic Society and its Useful Opposition.* New York: Anchor Books, 1968.

Self, Peter. *Econocrats and the Policy Process: The Politics and Philosophy of Cost-Benefit Analysis.* London: The MacMillan Press, 1975.

______. *Administrative Theories and Politics.* London: George Allen and Unwin, 1972.

Sharing in Development, A Programme of Employment, Equity, and Growth for the Philippines. International Labour Office, Geneva, 1974.

Sicat, Gerardo P. *The Philippine Economy in the Mid-Seventies: Development for the New Society.* Manila: Department of Public Information, 1974.

______. *Economic Policy and the Philippine Development.* Quezon City: University of the Philippines Press, 1972.

______. *The New Economic Directions in the Philippines.* Manila: National Economic Development Authority, 1974.

______. *The Philippine Economy in the 1960's.* Quezon City: The Institute of Economic Development and Research, University of the Philippines, 1964.

______. *The Philippine Economy in the 1970's: Prospects and Problems of Development.* Quezon

City: The Institute of Economic Development and Research, University of the Philippines, 1972.

Simbulan, Dante. *A Study of the Socio-Economic Elite in Philippine Politics and Government.* Ph.D. Dissertation, Australian National University, 1965.

Steine, Edwin O. and Associates. *Public Administration in the Philippines.* Manila: Institute of Public Administration, University of the Philippines, 1956.

Tatad, Francisco S. (ed.). *The Ascent: Growth and Change in Philippine Society.* Manila: Department of Public Information, 1974.

Technocracy Study Course. New York: Technocracy, Inc., 1974.

Touraine, Allain. *The Post-Industrial Society.* New York: Random House, 1971.

Valdepeñas, Vicente B. Jr. and G.M. Bautista. *The Emergence of the Philippine Economy.* Manila: Papyrus Press, 1977.

Weidner, Edward W. *Development Administration in Asia.* Durham, N.C. Duke University Press, 1970.

Wolters, William. *Politics, Patronage and Class Conflict in Central Luzon.* Quezon City: New Day, 1984.

Wurfel, David. *Filipino Politics: Development and Decay.* Quezon City: Ateneo de Manila University Press, 1988.

ARTICLES, MONOGRAPHS

Agpalo, Remigio E. " The Organic-Hierarchical Paradigm and Politics in the Philippines." Professorial Lecture Series Monograph No.1. Quezon City: University of the Philippines Press, 1973.

Apter, David. "The Role of the New Scientific Elite and Scientific Ideology in Modernization," *Comparative Studies in Science and Society*, 1974.

Arcega, Victoria M. "Technocrats in the Philippine Rice Program: Middlemen in Masagana 99," *Philippine Journal of Public Administration*, April 1976.

Bodenheimer, Susan J. "The Ideology of Developmentalism: The American Paradigm-Surrogate for Latin American Studies." Sage Professional Paper in Comparative Politics 01-015. Beverly Hills and London: *Sage Publications*, 1971.

Calderon, Aurelio E. "American-Trained Technocrats: Preliminary Views and Trends," *Philippine Journal of American Studies*, April 1979.

Carbonell, Ma. Aurora A. "Assessment of New Participation Mechanisms," *Philippine Political Science Journal*, June–December 1977.

Corpus, Onofre D. "Liberty and Government in the New Society: An Intellectual Perspective," *Philippine Political Science Journal*, June 1976.

"Development Policy and Planning: The Philippine Experience." A Monograph: Public Administration Special Studies Series No. 1. Manila: CPA Research Team, University of the Philippines, 1973.

Drietsel, Hans Peter. "Social Science and the Problem of Rationality: Notes on the Sociology of Technocrats," *Politics and Society II*, Winter 1972.

Dubsky, Roman. "Max Weber and Development Administration: Conflicts and Identities in Administrative Thought—Toward a New Development Administration." Occasional Paper No. 72-9 (October 1979), College of Public Administration, University of the Philippines.

________________. "Institutionalizing Social Conduct and the New Society in the Philippines," *Philippine*

Journal of Public Administration, January and April (Part I and II), 1974.

de Guzman, Raul P. " Politics in the Philippines," *Philippine Journal of Public Administration*, July–October 1971.

__________ and Associates. " Citizen Participation and Decision-Making Under Martial Law Administration: A Search for a Viable Political System." Occasional Paper, College of Public Administration, University of Philippines, March 1976.

__________ *et al.* " Citizen Participation and Decision-Making Under Martial Law," *Philippine Journal of Public Administration*, January 1977.

Lande, Carl H. " Parties and Politics in the Philippines," *Asian Survey*, September 1968.

Lapitan, Antonio E. " Citizens Assemblies in the Philippines: Institution-Building Under Martial Rule," *Asian Profile*, August 1975.

Laquian, Aprodicio A. " Martial Law in the Philippines to Date," *Philippine Political Science Journal*, December 1975.

Liddle, R.W. " Comparing Political and Economic Change in Indonesia and the Philippines: Some Preliminary Considerations," ISEAS paper (Singapore, 1977), unpublished.

Lynch, Owen J. and Talbott Kirk, " Legal Responses to the Philippine Deforestation Crisis," *Journal of International Law and Politics* (New York University), Spring 1988.

McCallus, Joseph P. " The Myth of the New Filipino: Philippine Government Propaganda During the Early Years of Martial Law," *Philippine Quarterly of Culture and Society*, 17 (1989); 129–148.

Nowak, T.C. "The Philippines Before Martial Law: A Study in Politics and Administration," *American Political Science Review*, June 1977.

Ocampo, Romeo B. "Social Justice: An Essay on Philippine Social Ideology," *Philippine Journal of Public Administration*, January 1971.

________________. "Technocrats and Planning: Sketch and Exploration," *Philippine Journal of Public Administration*, July–October 1971.

________________. "Blueprint for Political Modernization: The Present as the Future?" Paper prepared for the Third (1979) Annual Social Science Conference at U.P. Los Baños. (Mimeo.)

"Overview of Development Prospects," *Philippine Development*, November 30, 1976.

Pancho, Arturo. "Local Government Planning and National Development Policy of Four ASEAN Nations," *Philippine Journal of Public Administration*, January 1971.

"Planning Toward the City of Man," *Philippine Development*, July 15, 1977.

Ridley, F.F. "French Technocracy and Comparative Government," *Political Studies*, February 1966.

Robson, Alan. "Martial Law in the Philippines and Its Aftermath," SSED Working Paper No. 6, The University of the South Pacific, Suva, Fiji, 1987.

Sicat, Gerardo P. "Economic Independence: The Dilemma of the Third World," *Journal of Philippine Development*, Second Semester 1976.

________________. "A Historical and Current Perspective of Philippine Economic Problems." Address delivered before the Philippine Economic Society, December 8, 1984.

____________. "Revolution Endorsing the Government's Social and Economic Development Plans," *Philippine Development*, July–October 1971.

Soriano, Emanuel V. "The Philippines in the Year 2000: Tentative Prospects." Professorial Lecture Series No. 5. Quezon City: University of the Philippines Press, 1974. (Mimeo.)

Stauffer, Robert B. "A Legislative Model of Political Development," *Philippine Journal of Public Administration*, January 1967.

____________. "The Marcos Regime: Failure of Transnational Developmentalism and Hegemony Building from Outside." University of Sydney, Research Monograph No. 23, 1985.

____________. "The Political Economy of Development: A Philippine Note," *Philippine Journal of Public Administration*, April 1972.

____________. "Philippine Corporatism: Sectoral Integration of the New Society," *Philippine Journal of Public Administration*, July 1976.

"Summary: Five-Year Philippine Development Plan, 1978–1987," *Philippine Development*, December 31, 1977.

Turner Mark (ed.), "Regime Change in the Philippines: The Legitimation of the Aquino Government," Canberra, Australian National University, 1987.

Weintraub, D. "Development and Modernization in the Philippines: The Problem of Change in the Context of Political Stability and Social Continuity." Beverly Hills, California, *Sage Publications*, 1973.

DOCUMENTS

Philippine Development Plans.

Presidential addresses, speeches, reports on the state of the nation, *The Civil Service Reporter* (a Civil Service Commission publication, 1955–1980.

World Bank Reports; ADB Reports

OTHER SOURCES

Newspapers: *Bulletin Today*; *Philippine Daily Express*; *The Times Journal*; *Manila Bulletin; Business Day.*

The Fookien Times Yearbooks

The South

Index

G

H

I

N

O

P

Q

R

S

T